Adaptations of Desert Organisms

Edited by J.L. Cloudsley-Thompson

Springer
Berlin
Heidelberg
New York
Barcelona
Budapest
Hong Kong
London
Milan
Paris
Santa Clara
Singapore
Tokyo

Volumes already published

Ecophysiology of the Camelidae and Desert Ruminants
By R.T. Wilson (1989)

Ecophysiology of Desert Arthropods and Reptiles
By J.L. Cloudsley-Thompson (1991)

Plant Nutrients in Desert Environments
By A. Day and K. Ludeke (1993)

Seed Germination in Desert Plants
By Y. Gutterman (1993)

Behavioural Adaptations of Desert Animals
By G. Costa (1995)

Invertebrates in Hot and Cold Arid Environments
By L. Sømme (1995)

Energetics of Desert Invertebrates
By H. Heatwole (1996)

Ecophysiology of Desert Birds
By G.L. Maclean (1996)

Plants of Desert Dunes
By A. Danin (1996)

Biotic Interactions in Arid Lands
By J.L. Cloudsley-Thompson (1996)

Structure-Function Relations of Warm Desert Plants
By A.C. Gibson (1996)

In preparation

Physiological Ecology of North American Desert Plants
By S.D. Smith, R.K. Monson, and J.E. Anderson (1997)

Ecophysiology of Small Desert Mammals
By A.A. Degen (1997)

Ecophysiology of Amphibians Inhabiting Xeric Environments
By M.R. Warburg (1997)

Homeostasis in Desert Reptiles
By S.D. Bradshaw (1997)

Arthur C. Gibson

Structure-Function Relations of Warm Desert Plants

With 75 Figures

Springer

Dr. Arthur Charles Gibson
Department of Biology
University of California, Los Angeles
Los Angeles, CA 90095-1606, USA

ISBN-13: 978-3-642-64638-6 e-ISBN-13: 978-3-642-60979-4
DOI: 10.1007/978-3-642-60979-4

Library of Congress Cataloging-in-Publication Data Gibson, Arthur C. Structure-Function relations of warm desert plants/Arthur C. Gibson. p. cm. — (Adaptations of desert organisms) Includes bibliographical references and index. ISBN-13: 978-3-642-64638-6 1. Desert plants — Adaptation. 2. Desert plants — Ecophysiology. 3. Desert plants — Anatomy. I. Title. II. Series. OK922.G53 1996 581.5'2652 — dc20

Softcover reprint of the hardcover 1st edition 1996

Cover design: Design & Production GmbH, Heidelberg

Typesetting: Thomson Press (India) Ltd., Madras

Spin: 10077994 31/3137/SPS - 5 4 3 2 1 0 - Printed on acid-free paper

Preface

Introductory biology textbooks cite plants of warm deserts, especially cacti, as easily understood examples of natural selection. Historically, these lessons to young biologists emphasize how a peculiar structural design relates to storing or conserving water, and *xerophyte* and *xeromorphic* are now standard terms used to describe plants adapted to dry environments. Numerous reviews on xerophytism have been sources for such generalizations, but after critically rereading those early reviews now we discover that the conclusions were drawn mostly from species of semiarid vegetation rather than from arid sites.

The field of physiological plant anatomy needs to take a long, hard, fresh look at desert plant tissues, to distinguish desert structural adaptations from those characterizing plants of semiarid habitats, and to reanalyze the physiological significance of such desert designs. The current monograph focuses on plants of lowland tropical and subtropical arid deserts – for this exercise separated from cool desert, dry scrubland, and steppe–and avoids describing species that would be more typical of rocky upland desert sites. By restricting analysis in this fashion, investigators will have the clearest illustrations of structural adaptations to hot, nonfreezing, lowland environments. What readers will soon realize is that many features formerly ascribed to plants of dry habitats are noticeably rare in typical lowland desert plants!

To clarify basic anatomical properties of desert plants, this book was restricted in coverage to vegetative adaptations, and, regrettably, had to exclude reproductive adaptations from flowering through seedling establishment. In this series, seed biology was already treated in *Seed Germination in Desert Plants* (Gutterman 1993).

Numerous individuals have generously provided me with field-collected specimens from warm deserts, although admittedly my own observations and those in the literature document less than 10% of the entire desert flora. The remainder forms a test for the numerous hypotheses favored in this treatment. Thanks are due for the important conceptual framework contributed to the field of physiological anatomy from key researchers studying ecophysiology of desert plants.

Los Angeles, California
June 1996

A.C. Gibson

Contents

1 Plant Life Forms of Warm Desert Climates

Deserts occur where the driest climates on earth prevail. If the arid zone is contained within the 200-mm isohyet, then the warm deserts can be identified as arid regions in the subtropics and tropics having mean monthly temperatures that exceed 20 °C in the hottest month and 9 °C in the coldest month, hence locations that experience few, if any, and very brief subzero episodes per year. In these habitats, potential evapotranspiration (PE) is exceedingly high (greater than 2800 mm yr^{-1} from a Class A evaporation pan), the index of aridity, as defined by Oberlander (1979), typically is greater than 85 with a year-round soil moisture deficit or with a moisture index of Thornthwaite (1948) less than -40, and the Bowen ratio (annual sums of sensible heat flux/latent heat flux) is 3–7 (Kessler 1985). Aboveground net primary productivity is low, and in extreme deserts, where rainfall may not occur for 12 or more consecutive months, vegetation is sparse or completely absent. Warm deserts as defined here fairly closely fit Zonobiome III (Arid), excluding most Zonoecotone II–III transitions, of Walter and Breckle (1984), and occupy more than one-tenth of the continental landmass (McGinnies 1979). Most narrow, maritime fog deserts also qualify as warm deserts (Amiran and Wilson 1973), although fog deserts tend to have narrower daily and annual monthly ranges of temperature (Goudie and Wilkinson 1977; Evenari et al. 1985, 1986). Both lowland inland and coastal warm deserts are inhabited by similar plant life forms, products of convergent evolution in response to the drought-prone climates.

1.1 Distribution and Climates of Warm Deserts

Warm deserts are found mostly between 15–32 °N and S latitude (Fig. 1.1), largely resulting from persistent, descending high-pressure cells with hot, dry air (Cloudsley-Thompson 1977; Shmida 1985). Largest is the Sahara Desert of northern Africa, extending from the Atlantic Ocean eastward to the Red Sea and comprising more than 8 500 000 km^2 of lowland landscapes that are bordered by semidesert or other semiarid habitats. In the Horn of eastern Africa occurs the torrid, equatorial Somali-Chalbi desert region. Southern Africa has the coastal, foggy Namib Desert adjacent to the inland warm desert portions of the southern Kalahari and northern Karoo (von Willert et al. 1992). In Eurasia

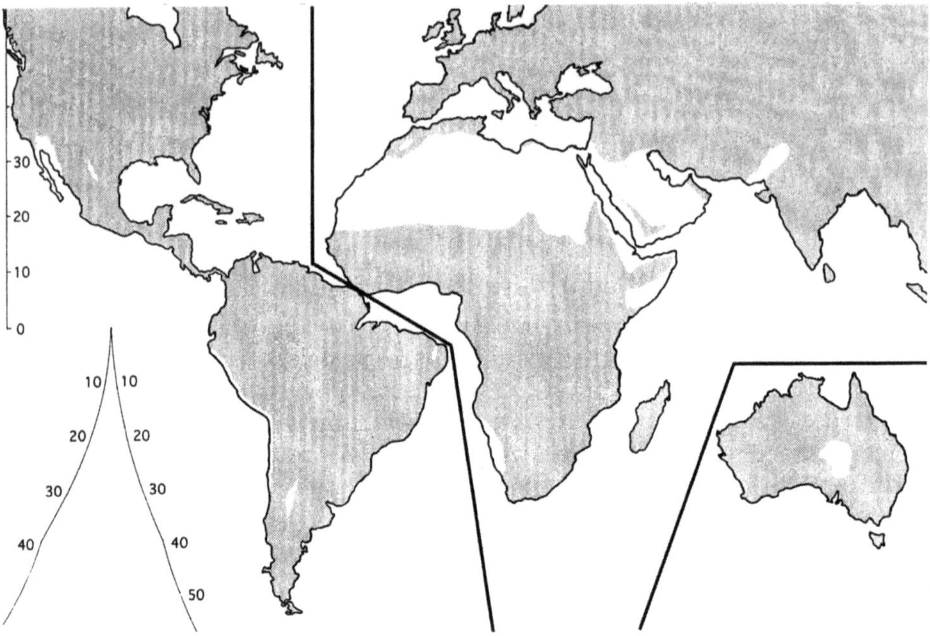

Fig. 1.1. Location of the warm deserts, as defined in the text

occurs the vast Arabian Desert on the Arabian Peninsula, to the north fringed by the cooler steppe desert of Iraq and the small rocky, warm desert of the Negev in southern Israel (Evenari et al. 1982). The Thar or Indian Desert of western India (Rajasthan) and adjacent Pakistan is a windy, sand-covered plain, apparently formed in the late Quaternary, and expanded through human activity (Bhandari 1978; Gupta 1986). An arid plain, called the Eremean Zone, constitutes the central one-third of Australia (Jessop 1981), but the driest portion (< 125 mm annual precipitation) is mostly the Simpson Desert, occurring north of Lake Eyre and southeast of Alice Springs in northern South Australia. Both the Thar and central Australia tend to have cool nighttime temperatures and some frost during winter. In the Western Hemisphere, the Sonoran, Chihuahuan, and Mojave deserts have arid regions, but many sites in the Mojave and Chihuahuan deserts may have snow and numerous freezing temperatures in winter. In South America, the best examples of warm deserts are northern Monte in central Argentina near San Juan and extremely arid portions of the Atacama and the equatorial Peruvian fog deserts along the western continental margin and low elevation slopes of the Andean Cordillera (Rauh 1985; Rundel et al. 1991). Several arid sites also occur on tropical islands and exposed coastal habitats.

1.2 Physiological Problems for Plants

With some regional exceptions, the typical warm desert environment is characterized by having: (1) low, variable, and often unpredictable precipitation, frequently resulting in drought during hot months; (2) high air and soil surface temperatures during part of the year or year-round; (3) low relative humidity and extremely high potential evapotranspiration; (4) high solar irradiation, often with cloudless conditions; (5) steady to strong gusty winds. These environmental conditions force perennial plants to survive in soils with limited availability of moisture, which thereby limits photosynthesis and productivity, and can place plant organs under lethal daytime thermal conditions, if they do not have adaptations for coping with high tissue temperatures.

If a desert region has a fairly well-defined rainy season, seeds of annuals are stimulated to germinate by the sudden increase in soil moisture. Although a rainfall event of 7 mm is generally sufficient to wet a soil to 0.2 m, wherein many fibrous roots of desert plants occur (Nobel 1976b; Rundel and Nobel 1991), studies have determined that one event or two closely successive rainfalls of 10 mm or less can induce germination of some species and an event of 20–25 mm is required for mass germination of ephemerals (Went 1948, 1949; Juhren et al. 1956; Koller 1956; Tevis 1958; Beatley 1974; Evenari 1985a; Inouye 1991). Extreme deserts, where rainfall is rare and unpredictable, may have an opportunistic flora of annuals following each rare rainstorm that produces enough water for germination. Annuals must survive soil drying at least long enough to form mature seeds and fruits.

Fog deserts rarely receive measurable rainfall, in some localities even for many consecutive years, but these deserts are special because precipitation from advective fog may exceed 150 mm yr^{-1}. Persistent banks of stratus clouds, several hundred meters in thickness, form just above a hyperarid littoral zone, and perennial plants of the fog zone provide a large surface area on which moisture condenses, producing fog drip sufficient to wet the upper soil horizon or, in some cases, to soil depths of 1 m or more (Ellenberg 1959). During the fog season, when ambient humidity typically exceeds 50%, plant growth occurs under the influence of fog when soil moisture is highest and PE is comparatively low (Seely 1978; von Willert et al. 1992). Dew condensation may be important in fog deserts but also independently supports life forms in the Negev (Evenari et al. 1982).

1.2.1 Water and Heat Stress

Plant water stress is a prime factor that inhibits the majority of mesophytes from developing natural populations in arid regions by affecting important physiological processes: (1) it limits cell and leaf expansion by decreasing turgor; (2) it causes daytime closure of stomata, thereby reducing not only transpiration but also photosynthesis, and increasing leaf temperature because heat dissipation is limited; (3) it stimulates leaf abscission when soils dry and leaf water content is reduced; (4) it apparently limits the photosynthetic pathway directly; (5) it causes air emboli to form and thereby block water transport in xylem conduits; and (6) it probably limits sugar transport in phloem sieve tubes. Native desert plants and successful invading weeds display a variety of ways to produce aboveground biomass when soil moisture is high and yet survive extreme soil water deficit with dormancy or at a comparatively reduced level of shoot productivity with measures to enhance water-use efficiency (WUE) while maintaining net carbon gain.

Water stress is severe year-round in extreme deserts or may occur during the coldest or hottest months, and life history strategies and functional morphology of warm desert plants are closely linked with seasonality of precipitation. Arid tropical and subtropical deserts may have predominantly late fall-winter precipitation, as in the westernmost Sonoran Desert and adjacent Mojave Desert, the southern Atacama, and Negev and northern Arabian Desert, or summer precipitation, as in the Thar, Somali-Chalbi, northern Monte, Chihuahuan, and northern Australian deserts. Broad deserts show regional differences in the rainy season. For example, in the Sahara, the northern portion receives most rainfall during cyclonic winter storms from the Mediterranean, and the southern portion during summer from the Intertropical Convergence Zone; rainfall in the central, extremely arid region is scanty, variable, and irregular, coming from either the north or south (Nicholson 1978). Another case is the central Sonoran Desert, which receives both winter and summer precipitation and therefore has a distinctive flora of annuals for each rainy season. Annuals, in particular, have evolved physiological mechanisms to maximize growth and reproduction for only one season (Mulroy and Rundel 1977).

For typical mesophytes, high temperature is deleterious because at chlorenchyma temperatures exceeding 46 °C heat stress impairs the thermal stability of cell membranes and proteins, probably causing leakage of ions (Pearcy et al. 1977; Berry and Björkman 1980; Taiz and Zeiger 1991). High tissue temperature greatly increases photorespiration and mitochondrial respiration relative to photosynthetic production, so that a plant can experience a net loss of carbon. Succulent plants with high heat capacitance and that lack mechanisms for cooling during peak daytime temperature can retain tissue viability even above 69 °C (Didden-Zopfy and Nobel 1982; Nobel and Smith 1983; Nobel 1984, 1989a; Smith et al. 1984; Kee and Nobel 1986; Nobel et al. 1986).

Hot dry winds increase transpiration when stomata are open, while they also intensify evaporation from the soil. Air with extremely low water content may also cause stomata to close (Schulze et al. 1974, 1975a,b; Bunce 1982, 1983, 1985, 1986; Sheriff 1984; Schulze 1986; Grantz 1990).

1.2.2 Soil Characters

The majority of warm desert soils contain a substantial portion of sand or sand and gravel and little or no organic matter, hence these soils have low water capacitance, dry relatively rapidly, and have limited available nitrogen (Fuller 1974; Evenari 1985a). Low soil nitrogen, as well as water, limits plant productivity. Highest local measurements of biomass have been recorded for desert woodlands of leguminous phreatophytes, such as species of *Prosopis*, *Acacia*, *Olneya*, and *Psorothamnus*, which tap groundwater or moist soil using deep roots and are hosts of nitrogen-fixing bacteria (Sharifi et al. 1982; Nilsen et al. 1984, 1991; Ludwig 1987). In extremely arid inland deserts, essentially all perennials grow only in *wadis* or *arroyos*, the ephemeral streams, washes, and runnels that channel floodwaters, because moisture can be found at one or more meters beneath the dry surface sand and gravel. Rock outcrops often provide moister or cooler microhabitats for plant species, around boulders and in rock crevices (Batanouny and Abdel Waheb 1973; Agami 1987; Nobel et al. 1992b) or on special rock types, such as gypsum, which has relatively high water capacitance (Meyer 1986).

Where the clay content is locally higher, e.g., swales between dunes or water catchment zones, plant growth may exceed that of plants on drier slopes. Playas, rich in silt and clay, often are saline or sodic, and can be colonized around the relatively nontoxic margins only by halophytes, especially Chenopodiaceae, Frankeniaceae, Tamaricaceae, and Poaceae. Halophytic populations may yield substantial standing biomass. A clay-rich soil also occurs in the relatively barren landscape of "desert pavement". The upper horizon is vesicular and consists mostly of sodic or saline-sodic clay, and water infiltration is extremely slow because deflocculation of soil colloids occurs and only a fraction of the precipitation infiltrates the jelly-like surface layer (Musick 1975; Dörn and Oberlander 1981). Much water of heavy precipitation events never enters the soil column for plant use (Evenari 1985a).

Soil horizons are weakly defined or absent in many warm desert aridosols. Stoney soils tend to be shallow, less than 1 m in depth, and frequently with a basement, rock-like horizon called calcrete, caliche, duripan, gypcrete, hardpan, kankar, or kundar. For example, calcrete is an indurated horizon of calcium carbonate and silica, located at a depth where water-carried calcium salts consistently penetrated year after year (Goudie and Pye 1985), and it is so hard that many roots cannot grow through the layer.

Pure sand on dunefields also lacks horizons, so roots in this substrate can grow unimpeded to reach moist sand one or more meters below the dune surface. Hence, stable portions of a dune may be colonized by species that require more water than those of more arid lowland desert scrub, e.g., White Sands in New Mexico (Emerson 1935; Shields 1951a; Dittmer 1959).

1.2.3 Wind and Flash-Flood Damage

Because warm deserts are regions of high pressure, they often experience strong to turbulent winds. Of course, hot, dry winds intensify evapotranspiration, but strong, gusty winds also buffet the tallest shoots. Unlike a mesic plant community, which has a plant canopy exerting frictional drag on wind near the ground, desert vegetation, which is composed of highly dispersed and low perennials, provides minimal form drag (Nobel 1991a), and thus is exposed to strong winds except close to the soil.

Strong, often turbulent prevailing winds are common in many arid regions and have shaped desert landscapes by forming ergs, huge areas of aeolian sand with massive dunefields. Ergs dominate more than one-third of the tropical and subtropical deserts, e.g., in Libya and Algeria in the Sahara, the Thar, Rub al Khali in Saudi Arabia, and the coastal Peruvian and Namib deserts. Wind abrasion from sandblast can damage aboveground plant organs. Shifting sand particles can either entomb or excavate entire plants, and wind-heaped aeolian deposits form "coppice mounds" or *nebkas* to cover the lower branches of perennials. Sandstorms can denude an area, and very few perennial species are capable of becoming established and surviving on active dunes. Moreover, during a sandstorm all flowers and young shoots shrivel due to the intensively dry conditions (Thomas 1921).

Deserts that receive monsoonal thunderstorms, and in particular those regions within or adjacent to mountain ranges, are subjected to sudden flash floods. Flash floods have huge discharges and velocities that may exceed 2 ms^{-1} (Fisher and Minckley 1978). Riparian species of wadis and arroyos must withstand the severe physical agitation of these infrequent but violent natural disasters.

1.3 Conceptual Models for Desert Plant Adaptations

Numerous authors have presented conceptual models for how desert plants respond to environmental stresses. These models typically contain either structural or physiological components or both, and therefore have been used in part for classifying desert life forms and life history strategies.

1.3.1 Xerophyte and Similar Terms

The Danish botanist Schouw (1823) is credited with introducing the term xerophyte to identify plants living in dry habitats, contrasted to plants from habitats with high water availability. Subsequent workers have defined xerophytes to include not only plants growing in arid and semiarid ecosystems but also those from wetter habitats that are well adapted to drought conditions, so-called physiological xerophytes (Lemée 1946), such as tropical epiphytes. However, some species from dry habitats show little tendency for water conservation (Kamerling 1914; Maximov 1929, 1931; Wood 1934) and thus were termed pseudo-xerophytes (Genkel 1946), whereas others are rainy-season ephemerals, which form shoots and can complete the reproductive phase before onset of severe water stress. For these reasons, Maximov (1931) redefined a xerophyte more narrowly as a plant growing in a dry habitat and that decreases shoot water loss under conditions of soil water deficit, often having a capacity to endure permanent wilting without injury.

Numerous ecologists have reviewed the pitfalls when utilizing a single term, such as xerophyte, and thereby generalizing about adaptation to arid climates (Delf 1915; Keller 1925; Maximov 1931; Thoday 1933; Shields 1950; Oppenheimer 1960; Seddon 1974). Unfortunately, researchers cannot assume that plants with "xeromorphic" features, i.e., structural adaptations for inhabiting dry conditions, use only water-conservation strategies, because results of early experiments showed that such plants can yield high transpiration rates when supplied with abundant water (Maximov 1931). Maximov correctly concluded that "the xerophytes are a very diversified group" and must be defined through showing "the real correlation between structure and function" and not simply one physiological or one structural criterion.

1.3.2 Stress Resistance

Desert plants display a set of strategies whereby an organism can deal with environmental stress, e.g., water stress during drought, a period without rains when soil moisture is inadequate to replace daily transpirational demands. One conceptual approach recognized several relationships of plants to drought (Kearney and Shantz 1911; Shantz 1927).

1. *Drought-escaping*—a plant grows during a short season and only where soil moisture conditions are highly favorable.
2. *Drought-evading*—a plant makes economical use of limited soil moisture, i.e., improved WUE), and can thereby evade early desiccation, or it can limit some portion of growth to lessen transpirational demand.
3. *Drought-enduring*—a plant is drought-deciduous and thereby drastically restricts water loss during the dry season.
4. *Drought-resistant*—a perennial plant can be physiologically active year-round, even when soil moisture is extremely low.

Although it has been widely used in desert plant descriptions, the model by Shantz has encountered difficulties because a given species may have several mechanisms to counteract water stress (Jones et al. 1981). Another criticism is that some species having very dissimilar strategies were lumped within a single category. Any unifactorial category disguises a fairly broad array of structural and physiological adaptations to deal with water stress, including species that appear to be drought-escapers, but actually can experience very low shoot water potentials (Ehleringer 1985b).

A simple conceptual model was used to classify plants growing in the Negev (Evenari et al. 1971; Danin 1983; Evenari 1985b). Plants growing in dry regions were described as (1) desiccation-tolerant, i.e., able to recover daily from dehydration; (2) plants active during the dry season (*arido-active*); and (3) plants inactive during the dry season (*arido-passive*).

Levitt (1972) introduced a formalized system to classify and study environmental stresses based on a mechanical engineering analogue. A biological stress is any factor that can induce a potentially injurious strain. Different stresses, such as water, temperature, pressure or bending, radiation, and chemical, can be analyzed independently. In his system, a biological response to a stress is termed *stress resistance* and may involve *avoidance*, i.e., partially or completely excluding the stress by means of a physical barrier, or *tolerance*, i.e., coming to a thermodynamic equilibrium with the stress without suffering injury, using either repair or prevention techniques.

According to Levitt, desert plants should have one of five strategies to respond to water shortage (Fig. 1.2). Plants that escape have no mechanisms to combat water stress and must complete the life cycle when ample free soil water is available. A plant resisting drought stress by avoidance attempts to maintain relatively high shoot water potential and turgid cells when exposed to dry environmental conditions; this includes "water-spenders", which must have timely water delivery to leaf surfaces to supply high daily water losses, and "water-savers", which show increased WUE as soil water potential decreases or other structural or physiological changes to decrease daily transpirational losses while less water is taken up from the soil. Levitt determined that drought avoidance is largely accomplished by structural modifications. A plant resisting drought by tolerance must be able to maintain live cells at low tissue water

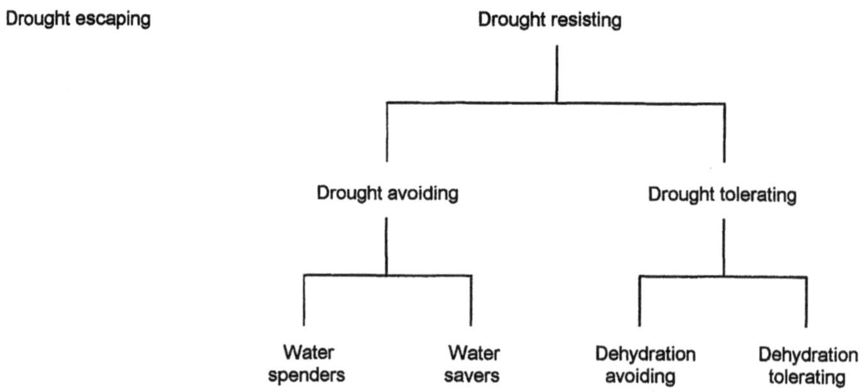

Fig. 1.2. Plant responses to drought stress. (After Levitt 1972)

content. Organisms with extreme desiccation-tolerant, or poikilohydric, photosynthetic structures, can be rapidly revitalized from a totally dehydrated state, i.e., so-called resurrection plants and lichens (Gaff 1977; Bewley 1979; Eickmeier 1979, 1980, 1983; Kappen 1988), and certain perennials can maintain positive turgor potentials via osmotic adjustment and thereby prevent stomatal closure when soil moisture would normally become limiting (Bennert and Mooney 1979; Turner and Jones 1980; Monson and Smith 1982; Calkin and Pearcy 1984; Nilsen et al. 1984). As suggested above, for coping with water stress a plant may have both avoidance and tolerance mechanisms.

To resist heat stress, desert plants generally employ avoidance, primarily by transpirational cooling, changed levels of radiant energy absorption, and insulation, but desert plants may also show heat tolerance and thus low sensitivity to high temperature due to protoplasmic resistance (Levitt 1980). Warm desert plants can be defined as those species that either lack tolerance of protoplasmic freezing or have structural mechanisms that enable the plant to keep shoot temperatures warmer than air during the short intervals of freezing nighttime temperatures.

Resistance to salt stress typically involves ATP-driven excretion of excess minerals with or without increased protoplasmic tolerance to higher osmotic concentrations. Similar mechanisms for excluding salts from leaf protoplasts occur in plants of warm and cool deserts as well as nondesert alkaline and saline habitats (Poljakoff-Mayber and Gale 1975; Osmond et al. 1980; Sen and Rajpurohit 1982; Staples and Toenniessen 1984; Cheeseman 1988).

A similar but slightly altered version of plant responses to drought stress refers to drought resistance sensu Levitt as drought tolerance mechanisms (Turner 1979; Jones et al. 1981), and these mechanisms were defined as: (1) avoidance of plant water deficits (escaping, water conservation, and effective water uptake); (2) tolerance of plant water deficits (turgor maintenance, protective solutes, and special osmoregulatory pathways); and (3) efficiency mechanisms that optimize water utilization and water harvesting. In the system by Turner, a sharp distinction is not drawn between drought-escaping and drought-avoiding water-spenders, because these can be treated as two extremes in a continuum of plant responses.

For arido-active plants, Evenari et al. (1975) proposed a list of adaptive features that enable plants to cope with water and high temperature stresses (Table 1.1).

1.3.3 Carbon Gain Models

At least since Maximov (1931), desert ecologists have been cautioned to observe that any strategy for a plant to retain water by restricting stomatal transpiration probably leads to a concomitant decrease in carbon gain via photosynthesis. Recognition of this biological dilemma has resulted in

Table 1.1. Adaptive features of arido-active plants under drought conditions (Evenari et al. 1975)

1. Tendency to develop xeromorphic structures
2. High root-to-shoot ratio
3. Reduction of metabolically active surface
a) Small surface in relation to dry weight
b) Seasonal surface reduction as a function of water stress
c) Partial death
4. Capacity to tolerate high soil water stress
a) Tolerance of high internal water-saturation deficits
b) Capacity to create and tolerate low water potentials
c) Photosynthetic activity even at low osmotic potentials
5. Reduction of transpiration rate through morphological and anatomical changes
6. Sensitive stomatal regultion as a function of ambient conditions
a) Regulation by water stress
b) Regulation by temperature
c) Regulation by air humidity
7. Adaptation of gas exchange mechanisms to high temperature

development of several models by assuming that survivorship and reproductive success of desert plants is highly correlated with traits that maximize net carbon gain.

A simple version of a cost-benefit model suggested that particular growth forms may have selective advantage in desert environments by maximizing net carbon gain, annuals with highest photosynthetic rates for a short growing season, drought-deciduous (rain-green) shrubs with intermediate rates over a period of several months, and evergreen shrubs with relatively low rates year-round (Orians and Solbrig 1977a; Solbrig and Orians 1977). With any functional convergence model, investment in structure should cease when return on that investment drops below a return in alternative investments (Field 1991). This type of model may be wrong if selection for maximizing net carbon gain is less important than adaptations to optimize WUE.

A model similar to cost-benefit analyses has related net carbon gain to tolerance of environmental stress (Ehleringer 1985b). This model (Fig. 1.3) describes possible life history strategies as tradeoffs between net carbon gain and tolerance to environmental stress. Long-lived perennial shrubs have relatively low net carbon gain per unit time but are buffered against environmental stress, whereas annuals, at the other extreme, have high rates of net carbon gain but restrict their active growth to relatively short periods when soil moisture is high. Short-lived drought-deciduous shrubs are intermediate in response.

Another type of conceptual model analyzes possible relationships between photosynthetic adaptations and a seasonally fluctuating thermal environment (Fig. 1.4; Mooney 1980b). Some plants can acclimate the thermal optima of photosynthetic organs to more closely match the daily temperature pattern encountered (Strain and Chase 1966; Mooney et al. 1978; Mooney 1980b; Ehleringer 1985b); some produce photosynthetic organs that are suitable for

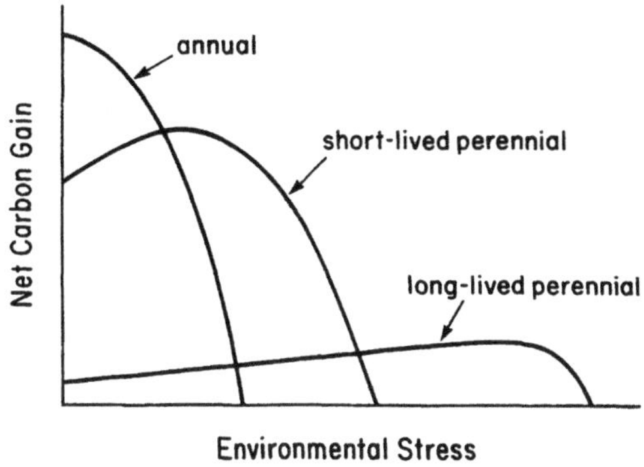

Fig. 1.3. Model of plant strategies relating net carbon gain to tolerance of environmental stress. (Rundel and Gibson 1996, after Ehleringer 1985b)

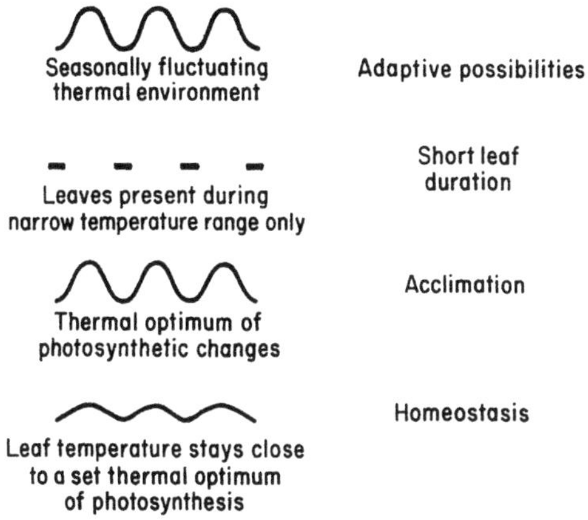

Fig. 1.4. Ecophysiological model showing relationships between photosynthetic adaptations and a seasonally fluctuating thermal environment. (Rundel and Gibson 1996, after Mooney 1980b)

only a narrow temperature ranges, e.g., cool spring or hot summer conditions; and some maintain homeostasis of a photosynthetic organ regardless of changes in the thermal regime. Biochemical capacity for carbon gain should vary with availability of resources (Mooney and Gulmon 1979, 1982).

1.4 Plants of Warm Deserts

1.4.1 Floras and Endemism

Detailed floras for many warm desert regions have been published, but vascular plants of vast areas are unsurveyed, so that only crude estimates for native and naturalized or introduced species are available (Shmida 1985). Widely disjunct desert zones have taxonomically very distinct floras, and even broad arid zones, e.g., the Saharo-Arabian and North American deserts, which have several widely distributed dominant species, have regional floras that contain endemic species (Shmida 1985). On each continent the dominant life form typically is taxonomically unique, showing that each flora was derived from different ancestors, and that life form as plants converged in adapting to similar environments (McGinnies et al. 1968).

1.4.2 Primary Vegetation Types

Each desert exhibits a set of primary vegetation types, governed to a large degree by physiognomy, climatic regime, and soil properties. No formal system has been employed to classify all types of warm desert vegetation, and workers have made relatively few quantitative comparisons between similar vegetation types on different continents (Walter and Breckle 1984; Shmida 1985).

Researchers in North America have classified typical glycophytic lowland vegetation as desert scrub (Barbour and Major 1977; Brown 1982). Along perennial and ephemeral desert washes and runoff runnels, desert scrub may be dissected by several subtypes of desert riparian woodland. Oasis woodland, which requires permanent water, may occur at desert springs. Other woodland communities are recognized where arborescent succulents, such as columnar Cactaceae, *Yucca*, and *Fouquieria columnaris*, are well established (Shreve and Wiggins 1964). North American desert grassland mostly occurs at upper elevation sites having relatively high precipitation values, hence semiarid climates. Low-lying alkaline habitats, which are colonized by halophytes, have been recognized as having distinct scrub communities (Barbour and Major 1977; Thorne 1982).

Monte in Argentina has plant communities similar to those found within the Sonoran Desert, and some of the dominant species are closely related to North American species (Orians and Solbrig 1977b; Brown and Gibson 1983).

In Old World deserts, arid regions support a population of very low-growing (to 0.2 m) dwarf shrubs and subshrubs (chamaephytes) that cover less than 10% of the total surface (Shmida 1985). For lowland chamaephytic communities, vegetation types have been identified using the dominant, indicator perennial, e.g., *Artemistietum herbae-albae*, *Zygophylletum dumos*, and *Zilletum spinosae* in the Middle East deserts. Huge lowland areas are extremely arid and therefore devoid of shrubs and trees (phanerophytes), except along wadis, this being comparable to desert riparian woodland or desert wash scrub (Thorne 1982). Oasis woodland communities are comparable to American cases, but, of course, contain different dominant species. Within sand-dune deserts, such as the Namib or Thar, plant associations have been described by location on or between dune faces (Robinson and Seely 1980).

Central Australia has unique aridland plant communities. The widespread, sandy lowland habitats are hummock grasslands dominated by perennial species of sclerophyllous spinifex bunchgrasses (*Triodia* and *Plectrachne*) (Winkworth 1967; Beard 1981). Even in the Simpson Desert, where rainfall is lowest, a desert grassland may support scattered shrubs and trees (phanerophytes), and species of riparian woodland are often those originating in semiarid mulga and mallee vegetation types. Sandy arid habitats of Australia typically lack leaf- and stem-succulent perennials, with the one widespread exception of *Sarcostemma viminale* subsp. *australe*, and sclerophylly is common here, whereas uncommon in other warm deserts. Treeless succulent steppe occurs primarily in a saline district south of the Simpson Desert, where xerohalophytes, especially Chenopodiaceae (e.g., *Atriplex nummularia, A. vesicaria,* and *Maireana sedifolia)* and *Nitraria billardieri* (Zygophyllaceae), are common chamaephytes.

Most difficult to classify conveniently are the varied vegetation types of the Peruvian-Atacama coastal desert. This contains vast stretches with absolutely no vegetation and regions with no perennials but rare rain-stimulated episodes of annuals to woodland dominated by arborescent cacti or legumes (Rauh 1985; Rundel et al. 1991). Along that 3700-km north-south fog desert, no zone is an exact physionomic or climatologic replica of another, nor are the many vegetation subtypes very similar to those in most other deserts, with some exceptions in west-coastal Baja California. Unique are the monospecific, psammophytic populations of terrestrial, rootless mats of *Tillandsia* species, which grow like beach strand communities, but within the fog belt.

1.4.3 Life Forms

Desert life forms have been summarized for many locations, but treatments have differed greatly on which categories and subcategories were accepted. Sometimes, categories have been adapted from Raunkiaer (1934; McGinnies et al. 1968): phanerophyte, chamaephyte, hemicryptophyte, geophyte, therophyte, and stem succulent. By itself, a Raunkiaer classification conveys few structure-function relationships as related to desert life. An elaboration of the Raunkiaer system, recognizing for desert taxa 16 subtypes, was published to relate life form to macroclimate, but again not specifically to arid habitats (Box 1981).

Shreve (1951; Shreve and Wiggins 1964) recognized 25 life forms for the vascular plants in the Sonoran Desert Region, and the examples were arranged as 14 broad categories (Table 1.2). By adding special life forms, such as desiccation-tolerant cryptogams (lichens, algae, mosses, and pteridophytes), stem and root parasites, and epiphytes, most of the warm desert plants can be conveniently categorized within this system, which, however, does not identify physiological parameters.

Zohary (1952) presented an informative phenological classification of woody desert plants of the Negev, taking into account water-use patterns of local species (Table 1.3).

From the standpoint of understanding structure-function relations, recent treatments have integrated knowledge of plant water relations, photosynthetic

Table 1.2. Classification of life forms and life history strategies in the Sonoran Desert (Shreve 1951; Shreve and Wiggins 1964)

1. Ephemerals (annuals) that were strictly seasonal versus those becoming facultative herbaceous perennials during very wet years
2. Perennial herbs with perennating belowground organs (geophytes)
3. Tussock grasses
4. Rosette plants with either succulent or nonsucculent leaves (e.g., acaulescent species of *Agave*, *Dudleya*, and *Yucca*)
5. Fruticose and arborescent monocotyledons with thick stems (e.g., tall species of *Yucca*)
6. Palms
7. Leafless stem succulents (e.g., Cactaceae)
8. Leaf succulents
9. Nonsucculent, aphyllous shrubs with photosynthetic stems
10. Deciduous shrubs with soft wood
11. Evergreen shrubs and trees with hard wood
12. Stout-stemmed drought-deciduous trees and shrubs (sarcophytes or pachycauls) with small or large leaves
13. Nonsucculent shrubs with photosynthetic stems and drought-deciduous leaves (e.g., *Cercidium*)
14. Winter-deciduous woody plants with relatively large leaves (includes phreatophytes)

Table 1.3. Hydroeconomic types of arid zones as defined by Zohary (1952)

1. Semideciduous dwarf shrubs experiencing drastic leaf surface reduction at the beginning of the dry season
   Examples: *Artemisia monosperma, Reaumuria palaestina, Salsola villosa, Suaeda palaestina, and Zygophyllum dumosum*
2. Heterophyllous dwarf shrubs having mesomorphic and xeromorphic leaf types
   Example: *Artemisia herba-alba*
3. Semievergreen virgate sbrubs having winter leaves and summer-green woody shoots
   Examples: *Calligonum comosum and Retama raetam*
4. Bud-evergreen dwarf shrubs that lose green winter shoots but retain evergreen buds
   Examples: *Reaumuria* spp.
5. Thorny, ephemeral-leaved dwarf shrubs having small early summer leaves, which are gradually lost
   Example: *Sarcopoterium spinosum*
6. Leafless semishrubs that lose some green shoots during summer
   Example: *Ephedra*
7. Leafless, modular semishrubs with green bark, shed from the previous year's bark during the summer
   Examples: *Anabasis articulata* and *Hammada scoparia*

pathways, and mechanisms of drought tolerance to identify typical desert plant strategies (Ehleringer 1985b, 1994; Smith and Nobel 1986), and these categories, with minor refinements, can be used to generalize for all warm deserts (Table 1.4).

Table 1.4. Classification of life forms of warm deserts used in the present book

1. Shrubs and subshrubs with drought-deciduous leaves. Woody perennials having drought-deciduous leaves as the principal photosynthetic organs, sometimes borne on succulent stems that may also experience some net carbon uptake
2. Evergreen shrubs. Woody shrubs with nonsucculent photosynthetic leaves year-round
3. Phreatophytes. Nonsucculent trees and shrubs that require roots to tap moist soil year-round
4. Perennial grasses. Tussock-forming Poaceae
5. Ephemerals. Annual and perennial seed plants forming aboveground shoots and flowers during a short period, often when soil water content is high, and dying aboveground when soil becomes dry
6. Nonsucculent aphyllous shrubs and trees. Perennials with nonsucculent stems as the principal photosynthetic organ and often with deep roots tapping substantial soil moisture
7. Stem and leaf succulents. Usually perennials having thick, water-storing tissues in the principal photosynthetic organ, but in others, photosynthesis mostly occurs in relatively large, deciduous leaves. These organisms resist desiccation when soil is dry
8. Poikilohydric cryptogams. Desiccation-tolerant organisms that withstand successive wetting and drying
9. Epiphytes and mat-forming air plants. Bromeliaceae with water-imbibing leaf trichomes and Crassulacean acid metabolism
10. Stem mistletoes. Perennial shoots present on woody stem of host plants
11. Xerohalophytes. Succulent to nonsucculent perennials and annuals living in saline soils and alkali sinks, and having special biochemical adaptations for salt stress

1.4.3.1 Shrubs and Subshrubs with Drought-Deciduous Leaves

In many lowland desert communities, prominent woody plants are C_3 species with drought-deciduous leaves. Most species have narrow leaves; if simple, often less than 5 mm wide and, if compound or pinnatifid, with narrow leaflets or blade segments, respectively. However, some very common shrubs, e.g., *Encelia farinosa* and *Hyptis emoryi* of California deserts, may have comparatively broad leaves. Shrubby phanerophytes and chamaephytes often are seasonally heteroblastic, producing wide mesomorphic leaves during spring, when water stress is low, and xeromorphic leaves as plant water potential decreases (Zohary 1952; Cunningham and Strain 1969; Ehleringer et al. 1976; Ehleringer 1976, 1982).

Among the nonsucculent woody perennials that during drought shed photosynthetic leaves, a significant proportion have long shoot-short shoot organization, whereby mature leaves can arise rapidly from buds along the length of a branch without appreciable stem growth (Fig. 1.5). In this way, a dormant desert-scrub community can produce a crop of photosynthesizing leaves within several days following an effective rainfall event (Humphrey 1975). Examples of long shoot-short shoot organization are *Lycium andersonii, Acacia neovernicosa, Porlieria angustifolia, Balsamocarpon brevifolium, Oxalis gigantea, Condalia globosa, Sericodes greggii, Tetracoccus hallii, and Phaulothamnus spinescens.*

Plants lacking long shoot-short shoot organization tend to shed branches at the onset of drought and greatly reduce leaf surface area (Zohary 1952). Probably for these and other deciduous woody perennials of deserts, leaf area index decreases as a function of seasonally declining plant water potential (Smith et al. 1995).

Fig. 1.5A–C. Long shoot-short shoot organization of desert shrubs. **A** *Porlieria angustifolia* of the Chihuahuan Desert. **B** *Euphorbia misera* of the Sonoran Desert; limber stems form thin leaves on young short shoots. **C** *Jatropha cardiophylla* of the Sonoran Desert; old short shoots have just begun to produce flower buds and leaf primordia

1.4.3.2 Evergreen Shrubs

Any evergreen desert species would be a true xerophyte. An evergreen retains physiologically active nonsucculent leaves during rainless months, i.e., it is drought-tolerant (Fig. 1.2). A list of common evergreen shrubs is short, but includes some very wide-ranging species, e.g., *Larrea tridentata*. Most desert regions have one or more evergreen species; of these, the majority display C_3 photosynthesis, but several species of *Atriplex* are C_4 evergreens. True evergreens, such as *L. cuneata*, may also experience great reduction in leaf surface during a dry season (Morello 1955); in fact, with the onset of drought, some so-called evergreens actually experience a loss of 70–90% of total leaf surface area and should be classified as drought-deciduous shrubs, e.g., *Artemisia herba-alba* and *Zygophyllum dumosum* of the Negev (Orshan 1963). The evergreen category includes phyllodineous species of *Acacia*, which often are arborescent.

1.4.3.3 Phreatophytes

Phreatophytes, which directly tap abundant surface and subsurface water supplies or moist soil, occur along perennial and ephemeral watercourses and in depressions where the stored water or water table is near the surface. According to Levitt (1980), phreatophytes are water-spenders (Fig. 1.2). In warm deserts, legumes are common microphyllous phreatophytic trees. *Prosopis* and *Acacia* are frequently seen in warm-desert riparian woodland, where they typically are deciduous species. However, *P. tamarugo* in the Atacama is evergreen (Mooney et al. 1980b; Rundel et al. 1991), and acacias of Australia have persistent phyllodes. In the Sonoran Desert, *Prosopis* may coexist in arroyas with the spring-deciduous *Olneya tesota*, essentially aphyllous *Psorothamnus spinosus*, leafy and green-barked *Cercidium* spp., or *Acacia* species, all legumes, as well as winter-deciduous *Chilopsis linearis* (Bignoniaceae). In the Saharo-Arabian and Thar deserts, rain-green, summer-active tree acacias coexist in wadis with the evergreen *Salvadora persica* or *S. oleifera*, and deciduous *Balanites aegyptiaca, Moringa aptera, and Ziziphus spina-christi.* Numerous shrubs also belong to these riparian woodland communities. Species of *Tamarix*, which are natives of drylands in Asia, also inhabit riparian habitats and are aggressive phreatophytes that frequently grow in monospecific stands.

Palms are rare in warm deserts and occur as natives only at permanent springs.

Northern Africa: *Hyphaene thebaica, H. repens, Medemia argun, Phoenix dactylifera* (introduced?), *Livistona (Wissmannia) churiensis, Nannorrhops stockeii.*
Central Australia: *Livistona mariae* (semiarid region).
North America: *Washingtonia filifera, W. robustus, Brahea brandegeei, B. armata, B. bella, B. decumbens, Sabal uresana.*
Monte: *Trithrinax campestris* (semiarid region).

These are shallow-rooted, broad-leaved evergreens that cannot survive without relatively high rates of daytime transpiration for evaporative cooling of leaves.

The wide, tough leaves of *Welwitschia mirabilis* of the Namib have relatively high transpirational losses (Cannon 1924; Schulze and Schulze 1976; Schulze et al. 1976b, 1980; von Willert et al. 1982; Winter and Schramm 1986) and probably use subsurface water.

In northern latitudes, several temperate genera with broad deciduous leaves, *Populus, Fraxinus*, and *Salix*, are found along desert streams and rivers (Liphschitz and Waisel 1970b). These species have very little resistance to water stress, and to exist in deserts must utilize groundwater; they share few of the structural properties of typical plants of desert washes and wadis, which have morphological and anatomical adaptations for desert life and can occur where they utilize deep moist soil, not groundwater.

1.4.3.4 Perennial Grasses

Perennial grasses intergrade with ephemeral grasses. Bunchgrasses, which are especially abundant in cooler steppe deserts, characterize vast regions of the

Australian arid zone (spinifex communities of *Triodia* spp.) and are the primary rhizomatous colonizers of sand dunes in Africa (e.g., *Stipagrostis*; Seely 1978; Robinson and Seely 1980). Species of *Pleuraphis, Achnatherum*, and *Bouteloua* are locally common in North American deserts. Prominent perennial grasses of warm deserts are most common where summer rainfall prevails and generally are C_4 species, although *P. rigida* begins growth in early spring of the winter-rainfall Mojave Desert (Nobel 1980d, 1981b).

1.4.3.5 Ephemerals

Ephemerals typically comprise a relatively small percentage of community biomass but a high percentage of total species in a desert flora. Included here are summer and winter annuals (therophytes), a few of which are facultative perennials (e.g., *Astragalus lentiformis*, Beatley 1970; *Fagonia arabica*, Bornkamm and Kehl 1989), spring- or summer-active geophytes, and assorted herbaceous perennials (hemicryptophytes), such as *Eriogonum inflatum, Mirabilis bigelovii, A. lentiginosus*, and *Tidestromia oblongifolia* from North American deserts. "Ephemeral" connotes a relatively short aboveground presence of leafy shoots, but the term can be misleading, in that some winter annuals may germinate in late fall and persist aboveground for more than 6 months (Beatley 1967; Mulroy and Rundel 1977; Inouye 1980, 1991; Inouye et al. 1980). Authors have described therophytes as drought-escaping or drought-avoiding, although some desert annuals may possess structural and physiological adaptations to improve rates of carbon assimilation and postpone water stress (Ehleringer 1985b). Winter annuals are C_3 species and summer ones mostly C_4 species (Mulroy and Rundel 1977; Ehleringer 1985b) unless, like *Dicoria canescens*, a summer-active C_3 annual has access to a constant water supply in dunes (Pavlik 1980; Toft and Pearcy 1982).

An ephemeral category should include spring- and summer-active herbaceous, geophytic, and low caudiciform vines, e.g., *Citrullus colocynthis* (Lange 1959; Althawadi and Grace 1986) and *Cucurbita palmata* (Bemis and Whitaker 1969; Rowley 1987); geophytic parasites on perennials, e.g., *Orobanche, Cistanche, Cynomorium coccineum*, and *Pholisma* (incl. *Ammobroma*); cryptic, endophytic stem parasites, e.g., *Pilostyles thurberi* on *Psorothamnus* and *Tristerix aphyllus* on *Trichocereus* (Mauseth et al. 1984); as well as annuals with succulent leaves, e.g., *Crassula connata*.

1.4.3.6 Nonsucculent Aphyllous Shrubs and Trees

Since Volkens (1887), descriptions of desert plants have included accounts on shrubs with old green, aphyllous, nonsucculent stems, which collectively function as the chief assimilatory organ of the plant (Gibson 1983). Each desert has representative taxa of the aphyllous green stem design (Chap. 4). Stem assimilants, known also as apophyllous xerophytes (Böcher and Lyshede 1968, 1972;

Böcher 1972), semievergreen virgate shrubs (Zohary 1952), and evergreen switch-branches (Stocker 1971), on adult shoots form photosynthetic microphylls or scales during rapid spring growth (Fig. 1.6A,B), but microphylls often abscise within a few weeks, and green stems continue plant photosynthesis for the remainder of the year. For some organisms, e.g., species of *Cercidium* (Adams et al. 1967; Adams and Strain 1968; Szarek and Woodhouse 1978a,b) and *Senna armata* in the Sonoran Desert and *Retama raetam* in the Sahara, the leaf or a portion thereof may persist for months as a C_3 photosynthetic organ; hence, the term aphyllous is not always an accurate description. Some species, e.g., *Justicia californica* and *Salazaria mexicana*, have leaves that are 10 mm or more in width and probably serve as major assimilating organs, and some species classified as deciduous shrubs, e.g., *Hymenoclea salsola*, have comparatively high rates of photosynthesis for both leaves and stems without periderm (Comstock and Ehleringer 1988).

Many aphyllous, green-stemmed species are facultative phreatophytes in that they most frequently are found along dry desert washes, where deep roots can obtain groundwater to remain active during drought months.

1.4.3.7 Stem and Leaf Succulents

Quite distinctive are plants with succulent photosynthetic organs. Storage of large water reserves in stem or leaves has long been considered one of the ob-

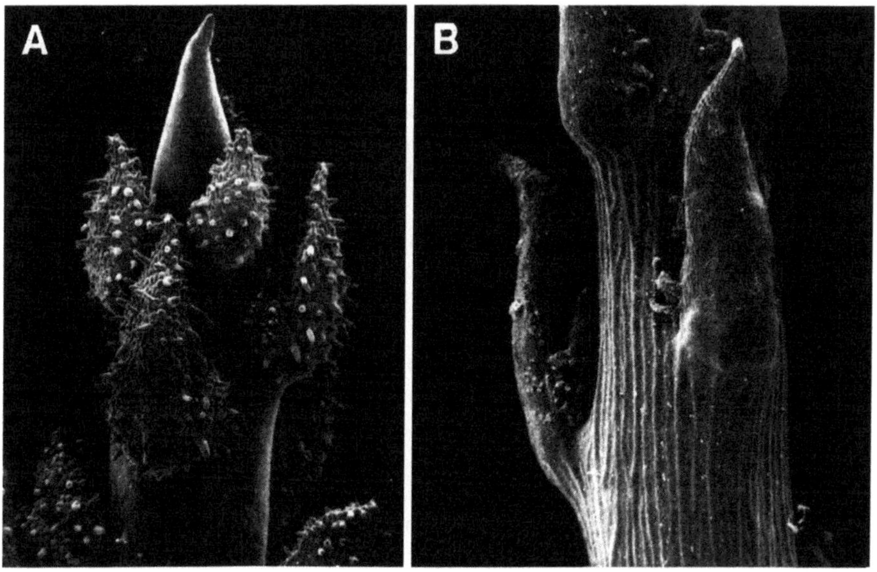

Fig. 1.6A,B. Scanning electron photomicrographs of young shoots from North American aphyllous, green-stemmed shrubs but here with vestigial, ephemeral leaves. **A** *Koeberlinia spinosa* **B** *Canotia holacantha*

vious strategies for desert life. This conclusion was reinforced for most warm desert succulents when Crassulacean acid metabolism (CAM) was explained as an adaptation to reduce transpiration by opening stomata during nighttime instead of the much hotter daytime (Gibson and Nobel 1986; Nobel 1988).

For a CAM stem succulent, the green stem is the chief, frequently the only photosynthetic organ, and stem succulents are especially characteristic of American deserts. Arborescent and fruticose columnar cacti are commonly the largest woody plants in desert communities where freezing rarely occurs, but many small cactus growth forms are also successful in deserts. A relatively small percentage of family Cactaceae (1550 spp.), less than 20% of the species, occurs in warm arid deserts, and cacti are more common in semiarid habitats, where rain regularly occurs at least once per year (Ellenberg 1981).

In African deserts, species of stem-succulent *Euphorbia* are morphological equivalents of New World cacti, but, like cacti, relatively few species of that genus inhabit arid zones; succulence in this genus has evolved mostly in semiarid habitats. In fact, the vast arid Sahara has no common stem succulents. Stem succulents of Stapelieae (Asclepiadaceae) are diversified in warm deserts of Africa, particularly in rocky habitats of southwestern Africa and the Somali-Chalbi Desert.

In North American deserts, *Agave* (Agavaceae) is the principal genus of CAM leaf succulents, whereas in Africa, CAM leaf succulents are extremely diversified, especially in Aizoaceae and Crassulaceae (von Willert et al. 1992). NonCAM leaf succulents are common with the fog belt of the Peruvian-Atacama deserts and South Africa. *Aloe* is an Old World genus of leaf succulents that has diversified mostly in semiarid habitats but occurs in many arid regions.

Yucca, which is generally a C_3 genus (Smith et al. 1983), includes important rosette species of the North American deserts, and several species are trees. Smith and Nobel (1986) treated yuccas as evergreen monocotyledons rather than leaf or stem succulents. *Yucca baccata* and *Y. torreyi* are CAM species of the Chihuahuan and eastern Sonoran deserts (Eickmeier 1978; Kemp and Gardetto 1982).

Conveniently treated as succulents is the heterogeneous assemblage of nonCAM caudiciform and pachycaul succulents. These plants have thick, water-storing stems or a stem-root axis (caudex) but typically form relatively broad, deciduous leaves, which are assumed to be the chief photosynthetic organs (Rowley 1987). For these, the thick stem or caudex may serve as a water reservoir, a buffer to permit extended maintenance of leaf turgor (Nilsen et al. 1990). In the Sonoran Desert, elephant trees (*Pachycormus discolor* and *Bursera microphylla*; Gibson 1981) and the boojum (*Fouquieria columnaris*) have chlorophyll-rich periderm but carbon assimilation of leaves predominates (Franco-Vizcaíno et al. 1990). Pachycaulous trees and shrubs in African deserts are *Cyphostemma currorii, Adenium obesum, Pachypodium namaquanum*, species of *Commiphora*, and *Moringa ovalifolia*. Succulent shrubs and subshrubs, e.g., *Adenia pechuelii* and *A. globosa*, which have a large caudex, when stimulated to grow, produce thin nonsucculent aerial shoots with deciduous leaves.

Other drought-deciduous shrubs, e.g., *Fouquieria* and *Jatropha*, have water-storing, succulent stems, and many of these have comparatively broad leaves, which are shed as soon as shoots experience mild water stress. *Fouquieria splendens* loses a crop of leaves when its shoot experiences water stress, but can rapidly produce a new cohort whenever the plant is rehydrated (Darrow 1943). Unlike nonsucculent shrubs and subshrubs, in these, stored water prevents stems from reaching low water potentials.

1.4.3.8 Poikilohydric Cryptogams

Desert cryptogams are poikilohydric, dehydration-tolerant organisms, capable of remaining metabolically competent following single or repeated bouts of cellular desiccation. Important physiological studies have been published on desert algae (Friedmann and Galun 1974; Evenari et al. 1982) and symbiontic lichens (Rogers 1977; Rundel 1978, 1988; Kappen and Rogers 1982; Kappen 1988), desert mosses (Nash et al. 1977; Bewley 1979; Rundel et al. 1979; Rundel and Lange 1980), resurrection-type species of *Selaginella* (Eickmeier 1979, 1980, 1982, 1983; Lebkuecher and Eickmeier 1990, 1991), and ferns (Nobel 1978a; Harten and Eickmeier 1987).

1.4.3.9 Epiphytes and Mat-Forming Air Plants

The genus *Tillandsia* (Bromeliaceae) includes species that have colonized fog deserts in Baja California and coastal South America. A few species are epiphytes, that perch on shrubs, but on the lomas of Peru and Chile thrive dense populations of several rootless caespitose terrestrial forms, *T. paleacea, T. werdermannii, T. latifolia,* and *T. landbeckii* (Rauh 1985; Rundel et al. 1991). These are CAM species that imbibe water through the upper epidermis of leaves via unique peltate trichomes (Benzing 1970; DeSanto et al. 1976; Rundel 1982).

1.4.3.10 Stem Mistletoes

Phanerophytes in deserts can be hosts to perennial mistletoe parasites (Viscaceae and Loranthaceae), which obtain water and nutrients from the plants (Schulze and Ehleringer 1984; Ehleringer and Schulze 1985; Ehleringer et al. 1985; Schulze et al. 1991).

1.4.4 Xerohalophytes

Widespread occurrence of saline warm-desert habitats, originating from long-term evaporation of storm runoff in low-lying, internal drainage basins, accounts for the appearance in desert floras of many halophytic genera, as well as cosmopolitan and wide-ranging species from salt marshes and other saline coastal communities (Shmida 1985). Xerohalophytes are a heterogeneous assemblage of succulent and nonsucculent organisms, not distinct as a separate life form, but differing in the ability to tolerate or avoid effects of high salt concentration on cell metabolism (Waisel 1972; Caldwell 1974; Poljakoff-Mayber and Gale 1975; Flowers et al. 1977; Osmond et al. 1980; Sen and Rajpurohit 1982; Munns et al. 1983; Staples and Toenniessen 1984; Cheeseman 1988). Interest here in halophytes as desert plants rests in anatomical and associated physiological techniques utilized to cope with salt stress, especially high levels of Na^+ and Cl^- in shoot tissues, and with heat and progressive water stress.

2 Functional Morphology of Nonsucculent Leaves

Aboveground green organs harvest sunlight and house all metabolic steps of photosynthesis, and they resist stresses with mechanisms to manage the energy budget and water relations of the entire plant. For many species of desert plants, where water is a major limiting factor, heat resistance is generally achieved by avoidance, which is largely morphological-anatomical in nature (Levitt 1980).

Unfavorable thermal regimes may result in death due to either high or low extremes, lowered photosynthetic capacity, and unprofitable respiratory costs. In particular, epidermal features play an important role in controlling water content of photosynthetic organs, set limits for carbon uptake and water loss, and ultimately determine longevity of aboveground organs and whole-plant survival. Thus, an integrated, multifactorial analysis of energy balance, shoot water relations, and primary productivity is required to determine the adaptational significance of each shoot design. Such analyses are somewhat different for nonsucculent plants (Chaps. 2–4), which have very little water or solar energy storage, and succulents (Chap. 5), which may have large organ capacitance for water and heat storage.

In general, it can be stated that external leaf characters affect energy fluxes, consequently leaf physiology, whereas internal structural modifications directly influence conductances in the gas diffusion pathway and the related photosynthetic capacity of the chlorenchyma. Biochemical mechanisms are primarily responsible for protoplasmic tolerance of water and temperature stresses (Björkman et al. 1980; Levitt 1980).

2.1 Energy Balance of a Leaf

Heat resistance by photosynthetic organs requires a balance between heat influx due to absorbed solar irradiation and environmental infrared radiation (reflected by surroundings) and heat efflux via transpiration, convection, conduction, and emitted infrared radiation (Nobel 1991a; Jones 1992). The energy balance of a leaf can be summarized as follows:

$$R_{abs} - L_e - H - \lambda E - q + M = 0 \tag{2.1}$$

where R_{abs} is absorbed solar irradiation plus absorbed infrared irradiation from surroundings, L_e is emitted infrared radiation, H is sensible heat exchange via conduction and convection, λ is the latent heat of vaporization for water with E evaporation rate, q is the heat representing storage, and M is energy storage or release by photosynthesis and other metabolic reactions (Gates 1970, 1980; Nobel 1974b). Photosynthetic pigments absorb in the visible region of 0.4 to 0.7 μm (PPFD, photosynthetical photon flux density, or PAR, photosynthetically active radiation, in μmol m^{-2} s^{-1}).

2.1.1 Absorptance as Influenced by Organ Orientation and Reflectance

Photosynthetic capacity and leaf temperature are strongly affected by the quality and quantity of absorbed solar irradiation, determined by (1) angle and location within the canopy of the photosynthetic organ relative to the direction of the sun's rays, and (2) reflectance of intercepted light energy. Values of leaf area index for most desert plants, which have small leaves and light-penetrating canopy designs, are low relative to those of other ecosystems (Chew and Chew 1965; Whittaker and Niering 1975; Cannell 1982). This allows many leaves to receive full-sun irradiance during a portion of the day.

Whenever the sky is dry, cloudless, and relatively dust-free, instantaneous absorbed direct solar irradiation will be maximal for a leaf oriented perpendicular to the sun's rays and minimal when it is oriented parallel to the sun's rays (Fig. 2.1). A species may have fixed leaf angles or suntracking leaves. Suntracking is a reversible leaf movement, achieved by turgor changes in leaf base or petiole, which allow a leaf or leaflet to remain perpendicular to the sun's rays throughout the daytime (diaheliotropic; Fig. 2.2A) or parallel to them (paraheliotropic; Wainwright 1977; Mooney and Ehleringer 1978; Ehleringer and Forseth 1980; Ehleringer 1985b; Ehleringer and Werk 1986; Koller 1990). Diaheliotropism results in maximizing PPFD, total diurnal exposure to direct solar irradiance, and net daily photosynthesis, but can also cause increased leaf temperatures. A leaf not perpendicular to the sun's rays may have lower heat

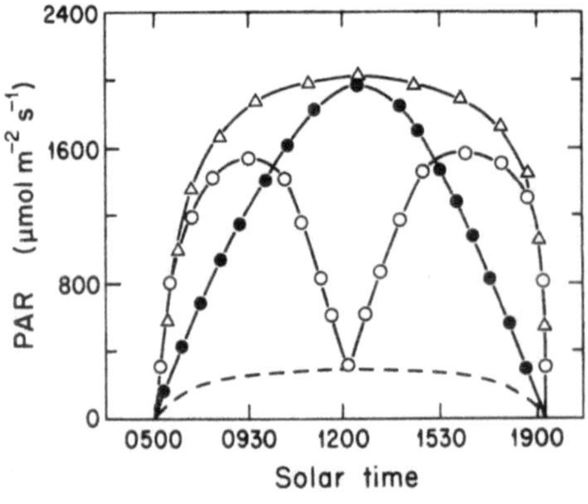

Fig. 2.1. Photosynthetically active radiation (PAR 0.4–0.7 μm) incident on four different leaves 1 m above the ground at a warm desert site near Palm Springs, California (33° 39′N) on a clear day of a summer solstice. Orientation of the lamina is measured fixed horizontal (planophilic, ●), measured fixed (east-west erectophilic, ○), simulated paraheliotropic (--), and simulated diaheliotropic (Δ). The diaheliotropic leaf would intercept 38% greater daily PPFD than either fixed leaf. (After Nobel 1980 and Ehleringer 1985b)

load, but will concomitantly have a lower photosynthetic rate because of reduced absorption of PPFD. Cupping, a movement of leaflets to become paraheliotropic, has been interpreted as a response to water stress (Wainwright 1977; Koller 1990). Elevated leaf temperature may be deleterious if it leads to less favorable water-use efficiency (WUE) and higher transpiration rates, or beneficial if it enables a leaf to operate closer to the thermal optimum for photosynthesis. Reflected infrared irradiation, which may constitute up to 30% of absorbed energy, can be received by a leaf at any angle and is strongly influenced by the albedo of and distance from soil and rocks (Nobel 1991a).

Leaves with fixed vertical or very steep angles (erectophilic; Fig. 2.2B) and with azimuth east-west orientation have been observed in deserts and semideserts (Mooney et al. 1977a, 1981; Werk and Ehleringer 1984). These leaves predictably receive most incident light in early morning and late afternoon and very little at midday (Fig. 2.1). Formerly, this condition had been simply interpreted as a strategy to reduce midday heating to levels tolerated by leaf metabolism; however, it may also represent an adaptation to maximize PPFD interception during periods, especially early morning, when leaf water potential generally is highest and air vapor pressure deficit is lowest. Bimodal diurnal illumination can be a more efficient daily use of PPFD, wherein a higher WUE is achieved by increasing photosynthesis with lower transpirational losses (Werk and Ehleringer 1984, 1986).

Whereas possible physiological effects of erectophily can be modeled and tested for single full-sun leaves, as well as for vertical cladodes (Nobel 1981a, 1982a, 1982b, 1986, 1988; Gibson and Nobel 1986), consequences of vertical leaf orientation have been extremely difficult to estimate for an entire canopy, such as an evergreen shrub with relatively high leaf area index. Self-shading may be associated with a variety of strategies to maximize certain daily or seasonal

Fig. 2.2A,B. A *Lupinus odoratus*, a diaheliotropic winter annual of California deserts with leaflets oriented toward the horizon just before sunset, maximizing PPFD. B Vertically oriented (erectophilic) leaves of the evergreen shrub *Leucophyllum frutescens* in southern Texas

physiological parameters of uppermost canopy leaves only or total canopy productivity as the azimuth of the sun varies over the course of the year (Nobel et al. 1993). In plants with significant self-shading, a complex algebraic analysis is necessary to determine which strategies predominate. Best studied has been canopy architecture of the evergreen *Larrea tridentata*, which in some sites displays southeasterly orientation of branches and east-west inclination of leaflets, apparently maximizing PPFD interception in early morning and late afternoon (Neufeld et al. 1988). Branch orientation and leaflet inclination appear to be adaptive for energy budgets of *Larrea* species in the Monte of Argentina (Ezcurra et al. 1991).

Upper canopy leaves of typical agricultural plants commonly absorb 75–85% of intercepted light, and many desert leaves have comparable values (Ehleringer 1981b). A leaf can avoid lethal temperature by reducing leaf absorptance of infrared radiation below typical green-leaf levels. This has resulted in the evolution of whitish, silvery, and grayish reflectant leaf types, best developed among perennials and achieved by epidermal modification. Certain types of dead trichomes appear to be adaptive to reflect light, and can reduce absorptance of light wavelengths from 0.4 to 3.0 μm (Ehleringer and Björkman 1978b), and some also reflect UV (Ehleringer and Cook 1987), which can damage important biological molecules (Caldwell 1981). Glaucous epicuticular wax may also be highly reflective, more especially for UV than for PPFD or heat-generating infrared radiation (Mulroy 1979).

2.1.2 Effects of Leaf Size and Form

Energy balances of leaves under hot and dry desert conditions have been modeled to understand evolutionary selection for leaf size and form (Gates 1968, 1980; Gates et al. 1968; Gates and Papian 1971). Broad leaves absorb much shortwave and infrared solar radiation, and as a consequence can greatly exceed ambient temperatures. Especially when the leaf-cell temperature exceeds 46 °C, which is a common midday summer temperature in warm deserts, several physiological processes may begin to break down (Björkman et al. 1980; Kappen 1981). To balance influx of absorbed radiation, heat must be dissipated by increased conduction, convection, or transpirational cooling [Eq. (2.1)]. The alternative, which has evolved in some desert plants, is to develop protoplasmic tolerance to high cell temperatures (Björkman et al. 1980; Levitt 1980). Energy conducted from the leaf or stem occurs across an unstirred air layer, the boundary layer, adjacent to the structure, through molecular or electronic collisions. Heat transferred from a leaf or stem is convected away outside the boundary layer. Conduction and convection may be strongly affected by boundary layer thickness, which is related to size and shape of the organ as well as to wind speeds. Transpiration, i.e., evaporation of water vapor from the plant surface, can produce a significant cooling effect from energy input requirements of vaporization, but the degree of cooling depends greatly on the size of the structure to be cooled and rate of water vapor loss from the surface.

Two computer simulations show how leaf size and transpiration rate very strongly affect midday leaf temperature, for when solar irradiance is high and constant (1000 W $m^{-2}s^{-1}$), ambient temperature is 40 °C, and relative humidity is 20%, as on a cloudless summer day (Fig. 2.3). In relatively still air (a wind velocity of 0.1 m s^{-1}), very broad leaves (200 mm width) with closed stomata may experience internal temperatures exceeding 55 °C, intermediate leaves (50 mm width) may have temperatures near 48 °C, and narrow leaves (10 mm width) may have temperatures near 44 °C (Fig. 2.3A). Temperatures of broad leaves can be reduced below lethal temperatures only by having very high rates of transpiration, whereas leaf temperatures of narrow leaves remain close to ambient temperature even without transpirational cooling. Under conditions with slightly greater wind velocity (1.0 m s^{-1}), nontranspiring broad leaves would still be close to lethal temperature at 48 °C, whereas narrow leaves would be below lethal conditions and within 1 °C of ambient temperature (Fig. 2.3B). Thus, the narrow leaf design is adaptive for hot desert climates by enabling a leaf to avoid lethal high temperatures without expending leaf water, merely by reducing boundary layer thickness and increasing the convective coefficient for heat loss.

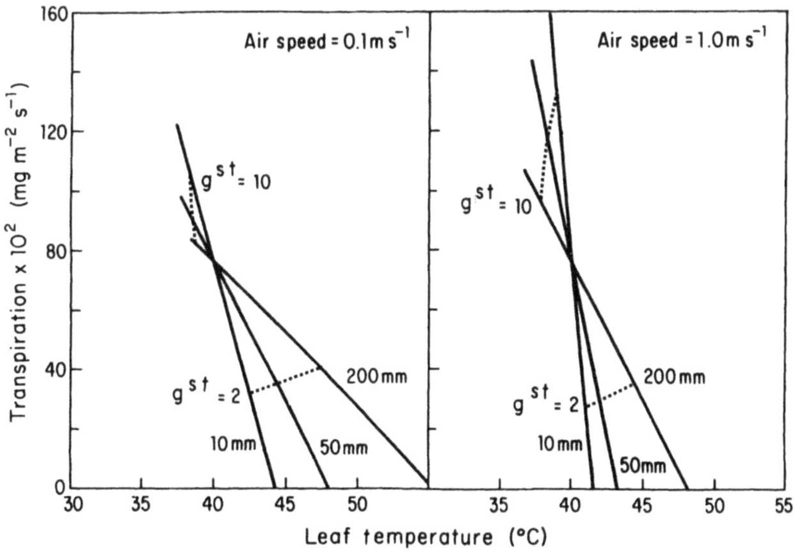

Fig. 2.3. Predicted temperatures of leaves 10 mm, 50 mm, and 200 mm in width (wind direction), computed from energy budget equations for still air (*left* 0.1 m s^{-1}) and a gentle wind (*right* 1.0 m s^{-1}) on a cloudless, summer day in a warm desert: solar irradiance 1000 W m^{-2} s^{-1}, ambient temperature 40 °C, and relatively humidity 20%. Stomatal conductances (g^{st}) of 2 mm s^{-1} and 10 mm s^{-1} are shown for each leaf width to demonstrate that at lower leaf temperature a given degree of stomatal opening will result in a substantially lower transpiration rate but experience no change in CO_2 uptake. (After Gates and Papian 1971)

For surviving summer drought, the cooler narrow leaf can be expected to have reduced transpirational loss. If internal leaf temperature is reduced by some degrees, then the temperature-dependent water vapor concentration within the leaf also decreases substantially, producing a smaller gradient to the dry atmosphere and hence a smaller driving force for evaporation from the leaf (Nobel 1991a).

A third adaptive advantage of a cooler narrow leaf is that, during hot days, a narrower leaf will operate closer to the thermal optimum for photosynthesis (often 25–30 °C) and at a reduced cost of respiration, maximizing carbon gain if photosynthesis occurs, and minimizing loss if it does not.

Models of leaf form have predicted that for two leaves of equal area, a lobed lamina should have greater convective coefficient for heat loss than the unlobed form, because the lobe is itself similar to narrower laminae, and compound leaves, with narrow leaflets, can remain cooler than simple leaves having equal leaf area (Balding and Cunningham 1976).

2.1.3 Energy Storage and Metabolic Heat Production

For relatively thin leaves, energy-storage terms can be ignored because they generate or lose a negligible quantity of heat, whereas thick structures, particularly those with heat-holding molecules, and especially succulents, can store considerable amounts of heat, which can require hours to dissipate. Metabolic heat production, especially as generated from respiration, can produce measureable, but not necessarily physiologically, significant amounts of energy in a warm desert leaf.

2.2 Leaf Size

2.2.1 Microphylly

It is universally recognized that typical nonsucculent desert leaves or leaf segments are small and narrow, i.e., microphyllous. Moreover, each desert species or desert clade characteristically displays narrower leaves than those of the sister taxon or population appearing outside the desert ecosystem (e.g., Mooney 1980a), evidence that strong selection has occurred for small and narrow leaves in arid habitats. Seedling leaves, formed when rapid growth is required for establishment and when soil moisture is high, generally are broader and have greater area than leaves formed on older, established plants, which must survive during the driest months; such plants are frequently aphyllous or have caducous, microphyllous leaves.

An analysis of 125 chamaephytes and low phanerophytes from the warm desert zone of California reveals that more than 89% of the nonphreatophytic, nonsucculent woody species have simple leaves, pinnatifid segments, leaflets, flattened rachises, or phyllodes less than 10 mm wide, and in most cases leaf segments are characteristically less than 5 mm wide (Fig. 2.4). Most of these woody perennials are either drought-deciduous or essentially aphyllous, with green photosynthetic stems. For all deserts, evergreen shrubs may have narrow leaves or leaf segments, e.g., in species of *Larrea, Maytenus, Atamisquea*, and *Mortonia*, whereas laminae wider than 10 mm occur in other evergreens of *Simmondsia, Capparis, Zygophyllum, Atriplex, Berberis*, and *Viscainoa.*

Early authors accepted a narrow-leaf design as a strategy to reduce the total transpiring surface area of the canopy, hence an adaptation to limit total water loss (Holm 1894; Cannon 1924). However, some studies have shown that total canopy leaf area of a desert plant may be comparable to or even greater than certain mesophytes of equal canopy height and cover, despite having small individual leaves (Ashby 1932). Nonetheless, leaf area per canopy area is comparatively low for desert shrubs, and it may be further reduced as a hydroeconomic strategy at the onset of drought, a phenology termed *shoot-shedding* (Zohary 1952).

Returning to energy budget parameters (Fig. 2.3), the most plausible explanation for microphylly in deserts is that leaf temperature can be maintained very close to ambient temperature without substantial transpirational cooling costs. This (1) prevents heat stress via avoidance and (2) reduces leaf temperature, which, in turn, results in lower transpiration rates (Gates et al. 1968). Field measurements of microphyllous desert leaves have consistently shown

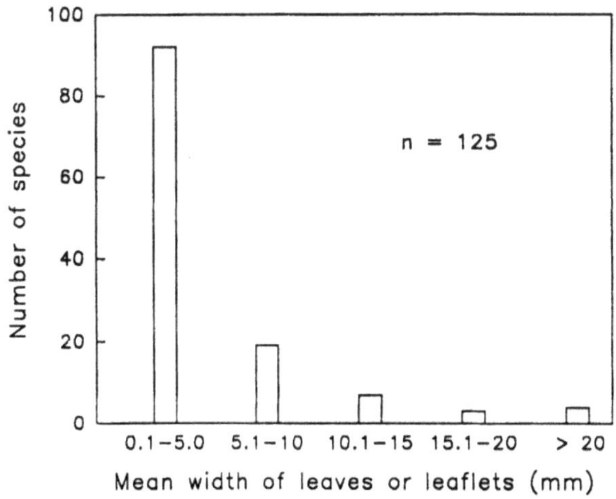

Fig. 2.4. Distribution of mean widths of leaves or leaflets from 125 native California species of woody desert perennials, showing that the majority have narrow or extremely narrow photosynthetic organs

that leaf overheating is typically less than 2 °C (e.g., Smith 1978). Hence, microphylly results in water conservation but not, as previously emphasized, by reducing only total canopy leaf area.

Microphyllous desert leaves tend to have little or no development of petioles and petiolules for leaflets. Absence of a petiole in a windy environment, such as a desert lowland habitat, would reduce torsional shear stresses on laminae (Niklas 1992) and would be expected to cause less interruption for bulk flow in xylem vessels and phloem sieve tubes at the base of the leaf.

2.2.2 *Broad Desert Leaves*

The models by Gates (Fig. 2.3A,B) predict that broad leaves may have remarkable over-temperatures when transpiration rates are low, but also may have under-temperatures when transpiration rates are extremely high. Lange (1959) reported a remarkable leaf cooling for *Citrullus colocynthis*, a broad-leaved desert vine in the Sahara, which, at midday in late summer and with a well-watered soil, had a temperature depression of 13 °C. Midday under-temperatures were confirmed for *Encelia farinosa, Eriogonum inflatum, Hyptis emoryi, Datura wrightii*, and *Sphaeralcea ambigua*, broad-leaved perennials of the Sonoran Desert, ranging from 8.4 to 18.1 °C, resulting from high transpiration rates between 6.2 and 14.3 mmol H_2O m^{-2} s^{-1}, which are values expected for well-watered mesophytes (Smith 1978). High transpiration rates in well-watered desert shrubs were appreciated by early physiologists (Maximov 1929, 1931; Wood 1934).

Examples of desert plants having broader leaves occur in most growth forms. Most are cases where actively growing shoots need abundant water to meet high transpiration losses, especially to reduce leaf temperature by evaporative cooling during a hot season (Fig. 2.3A,B). Evergreen dicotyledonous shrubs, which often have leaves greater than 10 mm wide, tend to have comparatively low transpiration rates, but, in addition, have evolved biochemical tolerance to high leaf temperature (Mooney and Gulmon 1979; Mooney 1980b).

Evergreen fan palms, such as *Washingtonia filifera* and *Brahea* spp., have the broadest leaves, and field studies have demonstrated high transpiration rates (Schmitt et al. 1993). For these leaves longitudinal splitting may reduce heat loads (Gates et al. 1968). For the Sahara, the fan palm *Hyphaene thebaica* had mesophytic rates up to 8 mmol H_2O m^{-2} s^{-1}, but experienced midday stomatal closure, whereas date palm (*Phoenix dactylifera*) had a much lower transpiration rate throughout the daytime (Ullmann 1985).

Balanites aegyptiaca, Salvadora persica, and *Ziziphus spina-christi* are three dicotyledonous trees of Old World desert wadis that often have relatively wide leaf surface (> 15 mm) and mesophyte-type transpiration rates (Stocker 1970; Sen et al. 1972). On the other hand, *Prosopis* spp., which have narrow leaflets, also have high transpiration rates (Nilsen et al. 1981, 1983, 1984). *Calotropis procera*, widely cultivated in arid lands, grows where plants can tap adequate soil moisture to supply extremely broad (100 mm) leaves. In desert riparian woodland and desert springs also are found phreatophytic trees of temperate genera, most notably *Populus* in North America and Egypt, which also have broad, mesophytic leaves.

Remarkable for desert floras is *Rumex hymenosepalus* of western North America, an arido-passive herbaceous perennial that may have leaves greater than 100 mm wide, and *Rheum palaestinum* (desert rhubarb) is a similar broad-leaved ephemeral of the Negev (Evenari et al. 1982; Danin 1983). Desert *Datura* are broad-leaved, arido-active forms. Some geophytes, such as *Jatropha macrorhiza*, appear during the summer rainy season of Mexican deserts. Other species of *Jatropha*, especially *J. cinerea* of Baja California and *J. cardiophylla* of the northern Sonoran Desert, have semisucculent stems with relatively broad leaves. Pachycauls and caudiciforms with relatively wide leaves include *Pachypodium* and *Adenium* of African deserts.

Cucurbitaceae in deserts require an abundant soil moisture supply for cooling relatively broad leaves. *Citrullus colocynthis, Cucurbita palmata, C. digitata*, and *Apodanthera undulata* are arido-active; each has a storage organ with high capacitance to provide some of the water for high transpiration rates and to produce high root pressures to refill shoot vessels (Lange 1959; Rundel and Franklin 1991), although roots still must tap soil water (Althawadi and Grace 1986; North 1992).

Endemic to the Namib is the gymnosperm *Welwitschia mirabilis*, which has very broad, tough leaves, a water-storing taproot, and a relatively high transpiration rate (Cannon 1924; von Willert et al. 1982; Walter 1986). These split basipetally into narrower segments sealed by wound periderm (Rodin 1958b; Salema 1967).

Even though most desert annuals are microphyllous, whether they grow predominantly in cool spring or hot summer months, the exceptions are fascinating. For example, winter annuals of southern California include many taxa with microphyllous leaves, e.g., species of *Nemacladus, Linanthus, Nama, Chaenactis*, and *Cryptantha*, which grow in open areas, and the widespread *Amsinckia tessellata*, which grows either in the open or within shrub canopies, and has comparatively broad leaves. In *Eriogonum*, with more than 25 species in the Mojave Desert, leaf width varies from a few millimeters in *E. pusillum* to more than 80 mm in late-spring and summer-flowering *E. brachypodum* and *E. deflexum*, which have thick, basal leaves that are cool to the touch on a hot afternoon, indicating high transpirational cooling. Coexisting with these species may be *Eremalche rotundifolia*, a broad-leaved diaheliotropic plant (Mooney and Ehleringer 1978), or *Geraea canescens*, a broad-leaved plant with fixed horizontal cauline leaves. Where summer rains occur, broad-leaved species of *Amaranthus* may occur with other taxa having narrow leaves.

2.2.3 Seasonal Heteroblasty

Numerous species of drought-deciduous woody perennials have heteromorphic leaves, relatively broad ones that develop during a season with high soil moisture, and narrow ones that are formed as soil moisture is depleted (Shreve 1924; Evenari 1938b; Orshan 1954, 1963; Zohary 1961; Cunningham and Strain 1969; Ehleringer and Björkman 1978b; Danin 1983). Perennials with distinctively heteromorphic leaves, e.g., *Encelia farinosa* and *Ambrosia dumosa* in the Sonoran Desert, *Artemisia herba-alba, Gymnocarpos fruticosus, Farsetia aegptiaca,* and *Phlomis brachyodon* in the Middle East and Sahara, form broad "green" leaves in cool weather and densely pubescent, narrower "white" leaves in hot weather. Heteromorphic leaves on adult plants have been described in California for the halophyte *Atriplex hymenelytra* (Mooney et al. 1977a) and phreatophytic species of *Prosopis* (Nilsen et al. 1986). Sun and shade leaves also occur in some desert perennials, as demonstrated by anatomical and biochemical measurements in *Hyptis emoryi* (Nobel 1976a).

Winter annuals commonly produce leaves in a basal rosette, which may be a strategy to improve heating during cool weather (Mulroy and Rundel 1977). Basal leaves tend to be broader and have greater laminal area than cauline leaves, and they typically wither first with the onset of soil water stress and increasingly hot soil surface temperatures, whereas the narrower or highly dissected cauline leaves persist as fruit set continues. In *Cymopterus longipes*, a herbaceous perennial, elevation of basal leaves at the onset of hot weather was thought to be a mechanism for avoiding heat stress (Werk and Ehleringer 1986). Reduced self-shading in winter annuals with basal rosettes may also contribute to rapid growth. Many annuals facultatively form broader leaves when soil moisture levels are relatively high and in narrower leaves in dry years and at dry sites.

Especially winter annuals having widely spaced microphyllous cauline leaves and wider leaves in a basal rosette, have canopy designs that produce little self-shading, and therefore probably contribute to rapid growth.

2.3 *Encelia* as a Model System

The genus *Encelia* has been used as a model system to evaluate the physiological importance of pubescence to the leaf energy budget and along aridity gradients (Cunningham and Strain 1969; Ehleringer et al. 1976, 1981; Ehleringer and Mooney 1978; Ehleringer and Björkman 1978a, 1978b; Ehleringer 1980, 1981a, 1982, 1983a, 1985a, 1985b, 1988; Ehleringer and Cook 1987, 1990). Several drought-deciduous species of *Encelia* are desert plants that exhibit leaf heteromorphy and produce smaller, heavily pubescent leaves in response to increasing water stress (Fig. 2.5). *Encelia farinosa*, a common chamaephyte of the Sonoran Desert, can produce "green," glabrescent leaves having up to 85% in-

Fig. 2.5. Dimorphic leaves of *Encelia farinosa*; the broad "green" leaf (*center*), produced in early spring when moisture was plentiful, and narrower "white," densely pubescent leaf (*right*) produced during late spring as the soil became dry at a California desert locality

tegrated PPFD absorptance varying in a graded series to late-season leaves having as little as 29% absorptance (Fig. 2.6A), the lowest measured value of any desert plant in North America (Ehleringer 1981a). In contrast, the sister species of *E. californica* in coastal sage scrub communities of California, has only green leaves with 85–88% absorptance. Along the coast of Chile, green-leaved populations of *E. canescens* occur in semiarid habitats, whereas white-leaved populations (48% PPFD reflectance) of *E. canescens* var. *lanuginosa* occur in extreme desert localities (Ehleringer et al. 1981).

Comparative studies of *E. farinosa* and *E. californica* have demonstrated that there are no internal physiological differences between them except light-dependence characteristics (Ehleringer and Björkman 1978a). Net photosynthesis is not light-saturated even under a clear sky in the summer (2000 μmol photons m^{-2} s^{-1}), and chlorophyll contents are equivalent (0.40 μg mm^{-2}), although a greater percent of chl_a occurs in *E. farinosa*. Photosynthetic rates (to 20 mg CO_2 m^{-2} s^{-1}) and CO_2 quantum yields per photon absorbed are relatively high when compared with other full-sun C_3 plants. Maximum water vapor conductance (to 20 mm s^{-1}) and internal CO_2 concentrations (250 μl l^{-1}) are similar. Neither species can acclimate to high temperatures (Strain and Chase 1966), which are probably lethal to both when greater than 45 °C. Both the mutant glabrate leaves of *E. farinosa* and white leaves with trichomes removed have absorptance curves nearly identical to those of *E. californica* (Fig. 2.6B), like the spectrum of a typical green leaf (Ehleringer and Mooney 1978; Ehleringer and Björkman 1978b; Ehleringer 1983a). Perhaps unexpectedly, *E. californica*

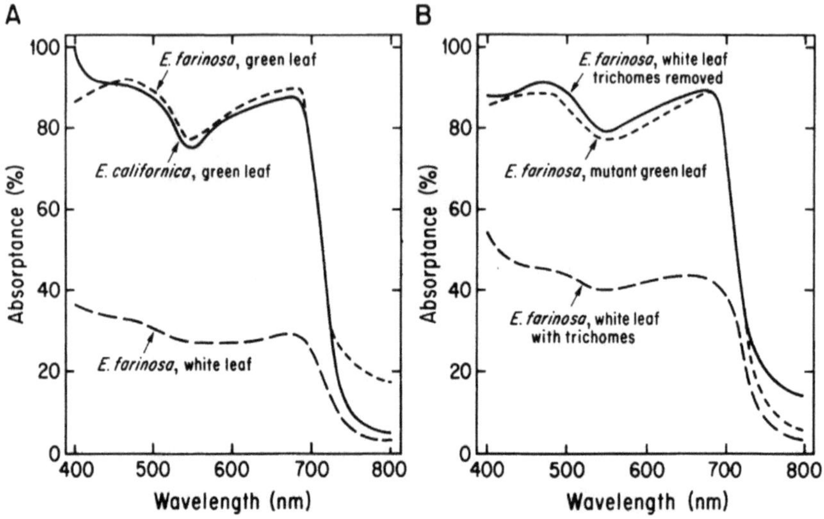

Fig. 2.6A,B. Absorption spectra of leaves over the wavelengths of 0.4–0.8 μm as determined with an Ulbricht integrating sphere. **A** Green leaves of semiarid coastal *Encelia californica*, and "green" and "white" leaves of the desert-dwelling *E. farinosa*. (After Ehleringer and Björkman 1978a). **B** From a different location, glabrescent mutant ("green") of *E. farinosa* from Death Valley, intact wild-type "white" forms having thick pubescence, and "white" form with trichomes removed. (After Ehleringer and Björkman 1978b; Ehleringer 1983a)

has approximately the same thermal optimum for photosynthesis as the desert species (30 °C). All this evidence suggests that *E. farinosa* invaded the desert environment by virtue of morphological, not biochemical or physiological, adaptations (Ehleringer and Björkman 1978b).

For *E. farinosa*, individual leaf area decreases and leaf thickness and reflectance increase steadily through a growing season as soil water is depleted (to −4.0 MPa) and air temperature increases (Smith and Nobel 1977b, 1978; Ehleringer 1980). Increased leaf thickness to 0.35 mm is due mostly to the growth of adaxial and abaxial mats of long, densely arranged, uniseriate trichomes, which are dead and air-filled (Fig. 2.7; Ehleringer and Cook 1987). Trichomes of *E. farinosa* act as a blanket reflector of infrared radiation (0.75 to 3.00 μm), profoundly reducing leaf temperature from expected values, by as much as 5 °C in summer at midday. At the same time, PPFD absorption is sharply decreased, so that photosynthesis, which is greatly light-limited, must occur at much reduced intensities, resulting in significantly lower carbon gain for densely pubescent leaves of *E. farinosa*. Hence, a trade-off exists at all light intensities in which carbon gain is sacrificed to avoid potentially lethal leaf temperatures.

Production of broad green leaves during cool weeks of the growing season and smaller white pubescent leaves during hot periods benefits plant carbon gain by optimizing leaf temperatures (Ehleringer and Mooney 1978). Winter

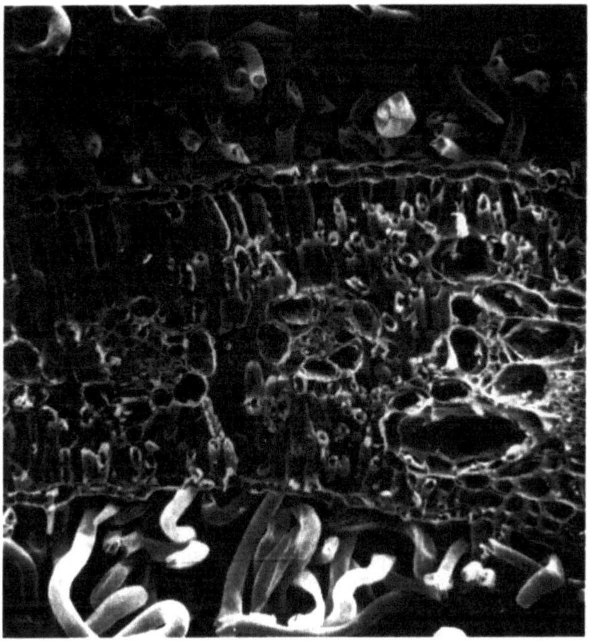

Fig. 2.7. Leaf transection of *Encelia farinosa* "white" (water-stressed) form, having relatively dense cover of twisted trichomes on both surfaces. Mesophyll is isolateral

and early spring leaves have over-temperatures up to 2.5 °C, bringing the leaf temperature closer to the photosynthetic optimum (T_{opt}; Fig. 2.8). In late spring and early summer, white leaves are cooler than comparable-sized green leaves, also moving leaf temperature downwards toward T_{opt}. Figure 2.8 shows how a summer leaf would attain 82% of the maximal photosynthetic rate at 37 °C, in comparison with 60% at 41 °C (predicted for a winter leaf). The leaf angle remains fixed at 26–28°, although a glabrate mutant in Death Valley had smaller leaves at angles twice as steep, and could only maintain the same leaf temperature as the wild type by maintaining higher transpiration rates (Ehleringer 1983a, 1985a, 1988).

As discussed earlier in the chapter (Sect. 2.1.2), one benefit to the smaller leaf is a lower leaf temperature during hot weather, which minimizes transpiration with substantial water conservation. In *E. farinosa*, a leaf temperature decrease from 41 to 37 °C on a summer day equals a 23% decrease in transpiration. Given that photosynthetic rate is improved by lowering summer leaf temperatures, WUE can increase significantly for the same amount of stomatal opening.

To interpret leaf heteroblasty in *Encelia*, as for other desert species, a combined leaf energy-photosynthetic model is required (Fig. 2.9; Ehleringer and Mooney 1978; Ehleringer 1980). In this model, certain types of dense, dead trichomes preferentially reduce IR absorption, thus decreasing leaf temperature to nonlethal and more productive levels, while also yielding lower transpira-

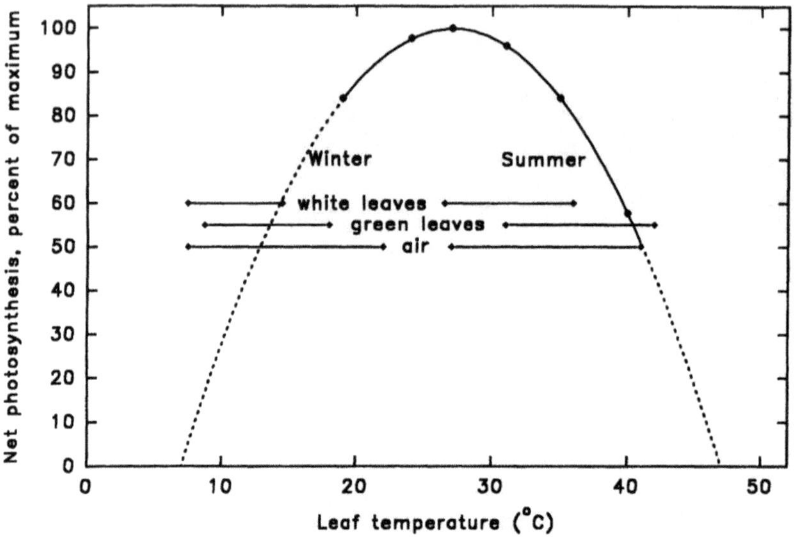

Fig. 2.8. Net photosynthetic rate of *Encelia farinosa* "green" winter and "white" summer leaves, plotted as percent of maximum over a range of leaf temperatures. *Horizontal lines with arrows* indicate ranges of field midday air temperatures and corresponding leaf temperatures for white and green leaf forms during winter and summer. (After Ehleringer and Mooney 1978)

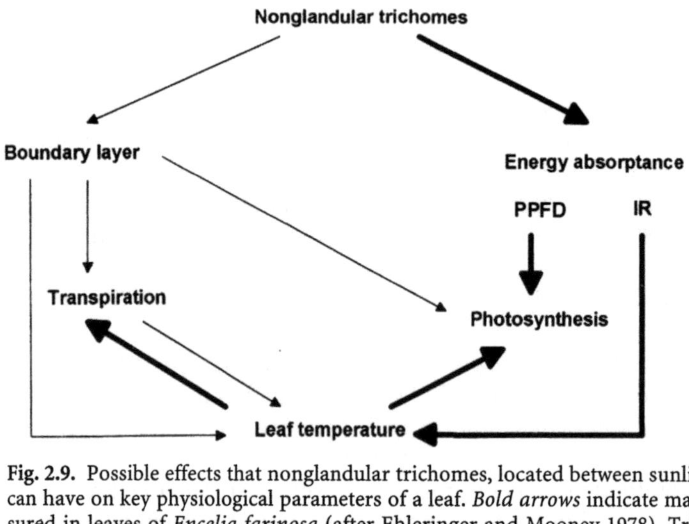

Fig. 2.9. Possible effects that nonglandular trichomes, located between sunlight and leaf tissues, can have on key physiological parameters of a leaf. *Bold arrows* indicate major impacts as measured in leaves of *Encelia farinosa* (after Ehleringer and Mooney 1978). Trichomes can reduce amount of PPFD and IR absorbed, and each has a different physiological consequence. In this example, effect on boundary layer was very small, but for a plant structure with an extremely thick, very dense trichome cover, physiological parameters would receive a larger impact via effect of boundary layer thickness

tion. Smaller leaves of drought months also may aid somewhat in lowering leaf temperature, although conductive cooling would be slowed slightly by having dense pubescence. Drought leaves of *E. farinosa* can remain functional at relatively low water potentials, and only wilt when midday values reach −6.5 MPa (Ehleringer 1982).

Unlike *E. farinosa*, which grows in typical *Larrea* desert-scrub communities, *E. frutescens*, which has smaller, short-lived, glabrescent leaves, inhabits washes and depends on greater transpirational cooling (Ehleringer 1983a; Comstock and Ehleringer 1984, 1986). High photosynthetic rates permit these plants to regain leaf carbon in as little as 7 days, and the rate remains above 80% maximal for air temperatures of 18–41 °C.

2.4 Structural Nature of Surface Reflectance

Reflectance from a plant surface can be modified by the presence of trichomes, cellular contours of the epidermis, or copious wax.

Trichomes occur on many desert leaves, and, in general, desert species often appear to have denser and thicker indumentum than sister taxa from semiarid habitats, although many opposite examples can be found. Trichomes occurred on 42% of xerophytes (Fahn and Cutler 1992); however, no study has determined whether occurrence of trichomes in deserts is statistically different than for species in semiarid plant communities. Indeed, numerous common perennials have glabrous to glabrescent leaves, suggesting that presence of trichomes is not a prerequisite for survival in warm desert ecosystems. Some trichomes shrivel or dehisce with age (Cannon 1924).

Trichome types are genetically determined and often show phylogenetic affinities, e.g., T-shaped or two-armed (many Brassicaceae) and L-shaped (Fabaceae), vesicular (Chenopodiaceae and Aizoaceae), stellate (Malvaceae and Euphorbiaceae), and dendritic (Lamiaceae).

As described above (Sect. 2.3), dead nonglandular trichomes may significantly reduce absorbed IR and thereby leaf temperature. North American desert species with highest reflectance (Ehleringer 1982) have relatively dense multicellular trichomes having several cellular designs (Figs. 2.10A–D). Highly reflectant trichomes are probably required for species lacking protoplasmic tolerance to high temperature, such as *Encelia farinosa*. They also occur on desert leaves that tolerate summertime temperatures greater than 50 °C, as in *Atriplex hymenelytra* (Pearcy and Harrison 1974; Mooney et al. 1975). Dead trichomes on most leaves are not thick enough to increase boundary layer thickness to a significant level, as was previously alleged (Volkens 1887; Johnson 1975; Fahn 1986).

Although certain nonsucculent leaves, notably in Fabaceae, tend to have well-developed epicuticular wax, copious surface wax is seldom observed on woody perennials, and where it occurs, loose wax is often greater on the adax-

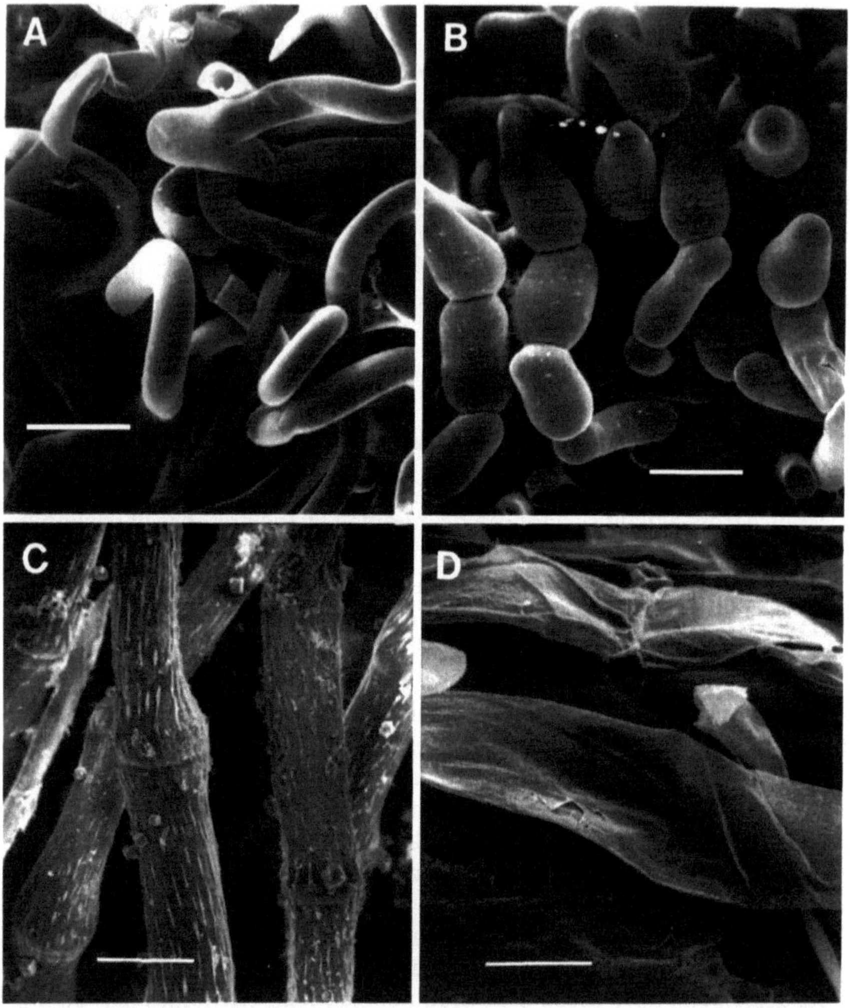

Fig. 2.10A–D. SEM photomicrographs of highly reflective trichomes on leaves. **A** *Encelia farinosa*. **B** *Psathyrotes ramosissima*. **C** *Dicoria canescens*. **D** *Pentzia incana*. Bar 25 μm

ial surface. In only a few cases are perennial leaves glaucous. Epicuticular wax produces the glaucous leaves of certain therophytes, especially in *Eschscholzia* and *Argemone* (Fig. 2.11A). Evergreen leaves of *Berberis trifoliolata* (Fig. 2.11B) and drought-deciduous leaves of *Becium burchellianum* are grayish, mostly due to a highly papillose epidermis (Fig. 2.11C). In contrast, glaucous stems and leaves of *Stephanomeria pauciflora* have a warty surface (Fig. 2.11D). Visually light leaves of desert *Lomatium* appear to be caused by folds or striae on epidermal cells. The importance, if any, of surface wax to the energy budget or UV absorption of nonsucculent desert leaves needs to be studied, because this fea-

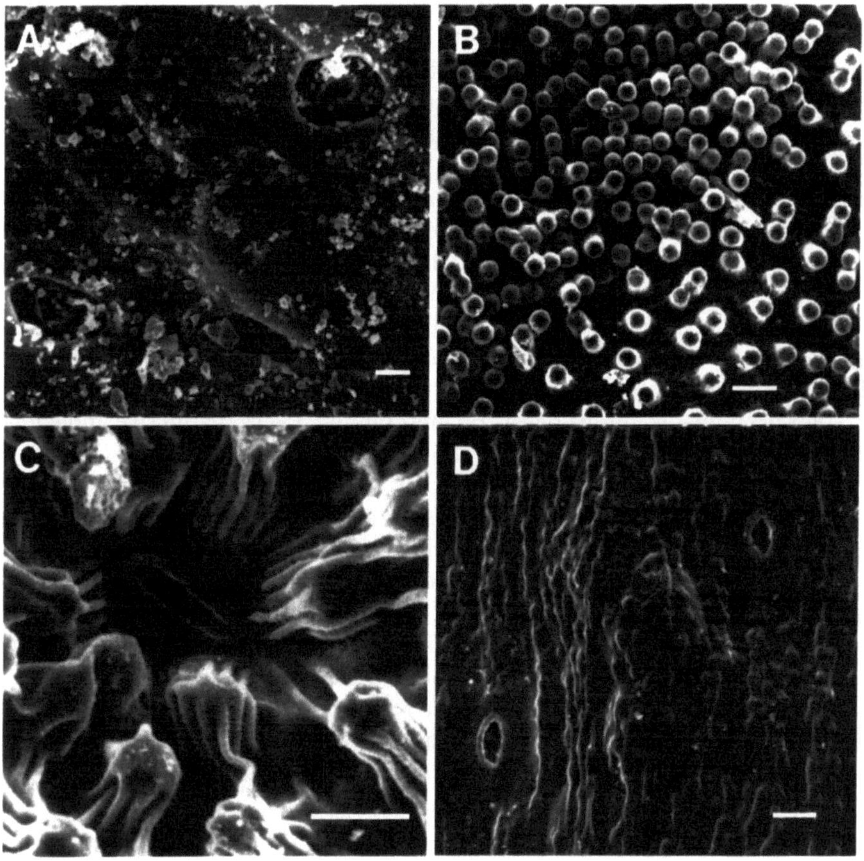

Fig. 2.11A–D. SEM photomicrographs of different types of reflective leaf epidermal surfaces. **A** *Argemone corymbosa* with epicuticular wax. **B** *Berberis trifoliolata* with prominent epidermal papillae. **C** *Becium burchellianum*, having epidermal papillae. **D** *Stephanomeria pauciflora* with a warty surface. Bar 50 μm

ture is much more common and probably more significant to the energy budgets of succulents.

2.5 Leaf Orientation and Display

Erectophily is most commonly observed among desert evergreens. In addition to the examples of *Larrea* discussed earlier in the chapter, other examples are *Mortonia sempervirens* and *Leucophyllum frutescens* of the Chihuahuan Desert, *Simmondsia chinensis* located along the moister margins of the Sonoran Desert, *Salvadora oleifera* in the Thar (Sen et al. 1972; Pillai and Pillai 1977), *Maytenus polyacantha* and *Zygophyllum prismatocarpum* in the northern Karoo, and *Atriplex hymenelytra* of washes and alkali sinks, especially in the Mojave Desert. The latter species is especially interesting because leaves are heteromorphic, being larger (> 500 mm^2) and less reflective in cool weather, and showing 70° leaf orientation during summer months (Mooney et al. 1977a). Steep leaf angles in *A. hymenelytra* result in greatly reduced leaf heat loads but little or no loss in carbon gain, because photosynthesis, which is limited at midday, is improved during early morning and late afternoon.

Diaheliotropism by desert plants has been demonstrated for numerous species but is particularly common among annuals and herbaceous perennials (Table 2.1; Ehleringer 1985b). This light-harvesting strategy has evolved in many different families and increases total daily photosynthetic rates (Mooney et al. 1976b, 1981).

For drought-deciduous desert phanerophytes, a very distinctive and common design for leaf display is long shoot-short shoot organization (Fig. 1.5), although the phenology of these plants has really not been studied very often, except for *Acacia raddiana* and *A. tortilis* (Halevy and Orshan 1973) and Mojave Desert species (Ackerman et al. 1980). Following a growth stimulus, a cohort of new leaves can be produced within several days from short shoots (brachyblasts), whereas growth of long shoots (dolichoblasts) may occur at a different time of the year. Brachyblasts and dolichoblasts are orthotropic shoots, and in some species, the stem becomes covered by sessile, brachyblastic leaves. In other cases, e.g., *Eriocephalus africanus* of the Karoo, the leaves of short shoots are minute, and give stems a tufted appearance.

Table 2.1. Native species of North American warm desert dicotyledons with heliotropic leaves (Mostly after Ehleringer and Forseth 1980)

Shrubs and trees
Acacia angustissima
Cercidium microphyllum
Prosopis juliflora
Psorothamnus emoryi
Senna bauhinioides
Herbaceous perennials
Abutilon parvulum
Anulocaulis annulatus
Astragalus lentiginosus
Boerhavia coccinea
Croton californica
Hymenothrix wrightii
Sida lepidota
Sphaeralcea ambigua
Annuals
Abronia villosa
Allionia incarnata
Amaranthus palmeri
Boerhavia wrightii
Chamaesyce abramsiana
Cleomella obtusifolia
Dicoria canescens
Eremalche rotundifolia
Eriogonum brachypodum
E. deflexum
Erodium texanum
Kallstroemia grandiflora
Lotus salsuginosus
Lupinus arizonicus
L. odoratus
L. shockleyi
Oxystylis lutea
Palafoxia arida
Proboscidea parviflora
Sphaeralcea coulteri
Tidestromia lanuginosa
Tiquilia nuttallii
Trianthema portulacastrum

Long shoot-short shoot organization has not been studied carefully from a functional point of view. Because lateral branching is suppressed, so that leaves form on fewer diverging axes, relatively high levels of PPFD penetrate these plant canopies, exposing many leaves to full sun for part of the day. From an energy standpoint, this shoot design, lacking internodes on its lateral shoots, can be interpreted as a saving of metabolites, which otherwise would have to be used for making woody stems. By eliminating most stem growth, leaves and flowers can form and mature very rapidly as soon as roots absorb water from unpredictable rains.

Factors potentially involved in the architecture of bunchgrasses, in terms of radiation interception, microclimate modification, nutrient cycling, and herbivory, has been reviewed recently (Smith et al. 1996).

2.6 Leaf Rolling and Revolute Margins

Some arid and semiarid grasses have mechanisms to change leaf form and physiology by leaf rolling during water stress Fig. 2.12). Many workers have suggested that leaf rolling is a xeromorphic adaptation to drought (Tschirch 1882; Beal 1886; Warming 1909; Weaver and Clements 1929; Shields 1951b; Lemée 1952, 1954; Oppenheimer 1960; McCleary 1968; Seddon 1974; Redmann 1985).

Poaceae are structurally predisposed for inrolling by having a leaf design characterized by parallel vascular bundles, reinforced by strands or girders of abaxial and adaxial fibers and often demarcated by intercostal furrows, and separated by flexible "hinges" of colorless cells extending between epidermal layers (Shields 1951b; Ellis 1976). Abaxial epidermis has stiff, thickened cell walls, so that when thin-walled cells of the adaxial epidermis, as well as other living and nonliving cells, shrink during water stress, laminal margins cup upward

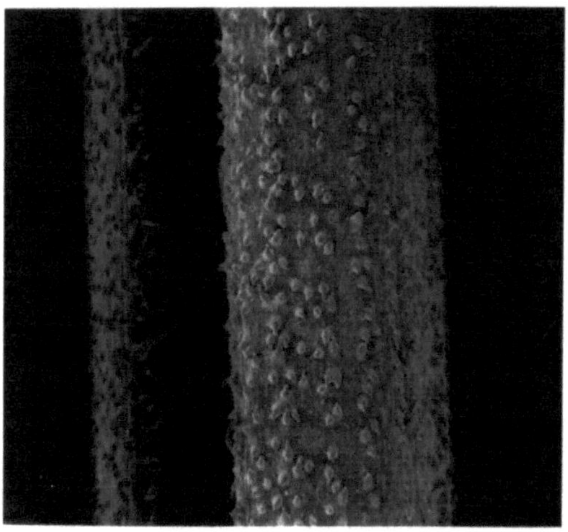

Fig. 2.12. Revolute leaf of *Achnatherum hymenoides*; as in many Poaceae, both sides of the leaf have axial rows of microhairs

and inward, becoming U-shaped, convolute, or involute, thus placing the abaxial surface on the outside. Leaf rolling is especially pronounced where large, vacuolate bulliform cells are positioned in the adaxial epidermis outside the hinges, and these cells buckle from passive compression.

Rolling of desert grass leaves can further reduce leaf width, already narrow (<10 mm), reducing leaf available to absorb radiation, and thereby affecting its energy budget. Leaf temperature would be expected to drop closer to air temperature, and intercepted PPFD in the internal mesophyll would also be reduced.

Leaf rolling has frequently been considered important for reducing transpiration, because inrolled leaves may experience a 46–83% reduction in transpiration (Oppenheimer 1960; Begg 1980). However, a recent analysis of flat versus rolled leaves of *Spartina pectinata*, a mesic C_4 grass with hyperstomatic leaves, suggested that decrease in transpiration was from a direct effect of stomatal closure, not changes in boundary layer conductance of the diffusion pathway or major temperature reduction; flat versus folded leaves of amphistomatic leaves from a different C_4 prairie grass showed similar results (Heckathorn and DeLucia 1991).

Energy budgets for leaf rolling of perennial desert grasses have not been analyzed. Within deserts are examples of leaves that are flat, thin, C_4, and amphistomatic (*Pleuraphis rigida*, Sonoran Desert); almost flat, somewhat thick, C_4, and hypostomatic (*Centropodia glauca*, Namib; Ellis 1984); U-shaped, thick, C_3, and hypostomatic (*Dregeochloa pumila*, Namib; Ellis 1977); convolute, thick, C_3, and amphistomatic (*Achnatherum hymenoides*, Mojave; Shields 1951b; Redmann 1985); and convolute, thin, C_4, and amphistomatic (*Stipagrostis sabulicola*, Namib; de Winter 1965), *S. pungens*, Sahara (Stocker 1972), and *Panicum turgidum*, Sahara.

In poikilohydric "resurrection" vascular cryptogams, rolling of fern fonds or entire photosynthetic shoots occurs during dehydration. When physiologically active, shoots of *Selaginella lepidophylla* form a flattened rosette, but during desiccation, the entire shoot curls tightly into a sphere. Curling reduces exposure of dehydrated photosynthetic tissues to high light and leaf temperature, and serves to limit damage by reducing photoinhibition and breakdown of the chloroplast photosystem apparatus. When curled, *S. lepidophylla* experiences a 99.7% light attenuation (Lebkuecher and Eickmeier 1990).

The resurrection-type rose of Jericho, *Anastatica hierochuntica*, displays a physical hygroscopic response of dead plants, which bear mature fruits (Steinbrinck and Schinz 1908; Evenari et al. 1971; Friedman et al. 1978).

Woody dicotyledons having leaves with revolute margins are rarely encountered in warm deserts, in contrast to their common occurrence in sclerophyllous scrub communities of mediterranean-type climates (Kummerow 1973) or cool and high elevation xerophytic plant communities (Böcher 1979). The best examples of revolute leaves are several salt-excreting xerohalophytic species of *Frankenia*, *Thymelaea hirsuta* (Stocker 1974a) that is also found in maquis vegetation, and the mostly cool-desert species *Coleogyne ramosissima*.

Several other species may occur in typical lowland desert scrub, but are actually typical residents of semiarid habitats that sometimes range into the arid zone. In these species, adaxial trichomes are mostly lacking, and abaxial indumentum may be involved in blocking IR irradiation reflected from soil or rock surfaces.

3 Physiological Anatomy of Nonsucculent Leaves

Many long-standing explanations for anatomical adaptations of desert plants and xerophytes can be traced to a classic analysis of Egyptian desert plants (Volkens 1887). This work proposed principles for desert botany and set the tone for explaining each anatomical feature of a desert photosynthetic organ as an adaptation primarily designed for water conservation. Volkens was correct to emphasize the gas diffusion pathway in his analysis, but overlooked the potential value of these anatomical features for determining rates of CO_2 diffusion. The modern approach, which must challenge many often-cited explanations in the literature, attempts to interpret the relative significance of each parameter affecting water vapor and CO_2 as independent fluxes with overlapping, but not identical, diffusion pathways (Nobel 1991a; Jones 1992).

3.1 Gas Diffusion Pathway

A gas diffuses into or out of a plant organ along a concentration gradient. Under typical daytime conditions, water vapor concentration is extremely high (air nearly saturated) within intercellular air spaces of a leaf (c_{wv}^{ias}) and low in air surrounding the leaf (c_{wv}^{a}); hence, an efflux of water vapor is expected. Conversely, CO_2 concentration in turbulent air ($c_{CO_2}^{\mathrm{a}}$) is approximately 360 ppm, whereas in chloroplasts, where CO_2 is consumed in the dark reaction, CO_2 concentration ($c_{CO_2}^{\mathrm{chl}}$) is considerably lower, hence an influx of CO_2 is expected. Differences in endpoint concentrations (Δc) of diffusion pathways are attributed to physical conditions in the atmosphere and to physiological processes within cells, but between these two endpoints are a series of structural parameters that greatly influence diffusion rates.

3.1.1 *Water Vapor*

Analyses of leaf water vapor loss have been conceptualized as a gas-phase diffusion pathway from H_2O-saturated pores in mesophyll cell walls and other interior surfaces of water evaporation through intercellular air spaces, stomata, cuticle, and across a boundary layer to turbulent air (Nobel 1983a, 1991a; Jones 1992). A similar series of barriers is found in stems from cortex to the atmosphere (for Chap. 4), sometimes with the added complexity of periderm components. Each component of the pathway imposes a resistance to diffusion (r in s mm^{-1} or s mmol^{-1}), analogous to resistance in electrical circuitry, which limits water vapor fluxes out of a leaf, or, conversely, the reciprocal term conductance ($g = 1/r$ in mm s^{-1} or mmol s^{-1}). Both terms are widely used, but here conductance is preferred because it is easier to visualize diffusion rates in those units.

In the electrical circuit analogy (Fig. 3.1A) water vapor conductance through the intercellular air space (g_{wv}^{ias}) and stomata (g_{wv}^{st}) is in series: $g_{wv}^{ias}\,g_{wv}^{st}/(g_{wv}^{ias}+g_{wv}^{st})$. Water vapor movement through cutinized outer epidermal cell walls (g_{wv}^{c}) is a parallel conductance:

$$g_{wv}^{leaf} = \frac{g_{wv}^{ias}\,g_{wv}^{st}}{g_{wv}^{ias}+g_{wv}^{st}} + g_{wv}^{c} \tag{3.1}$$

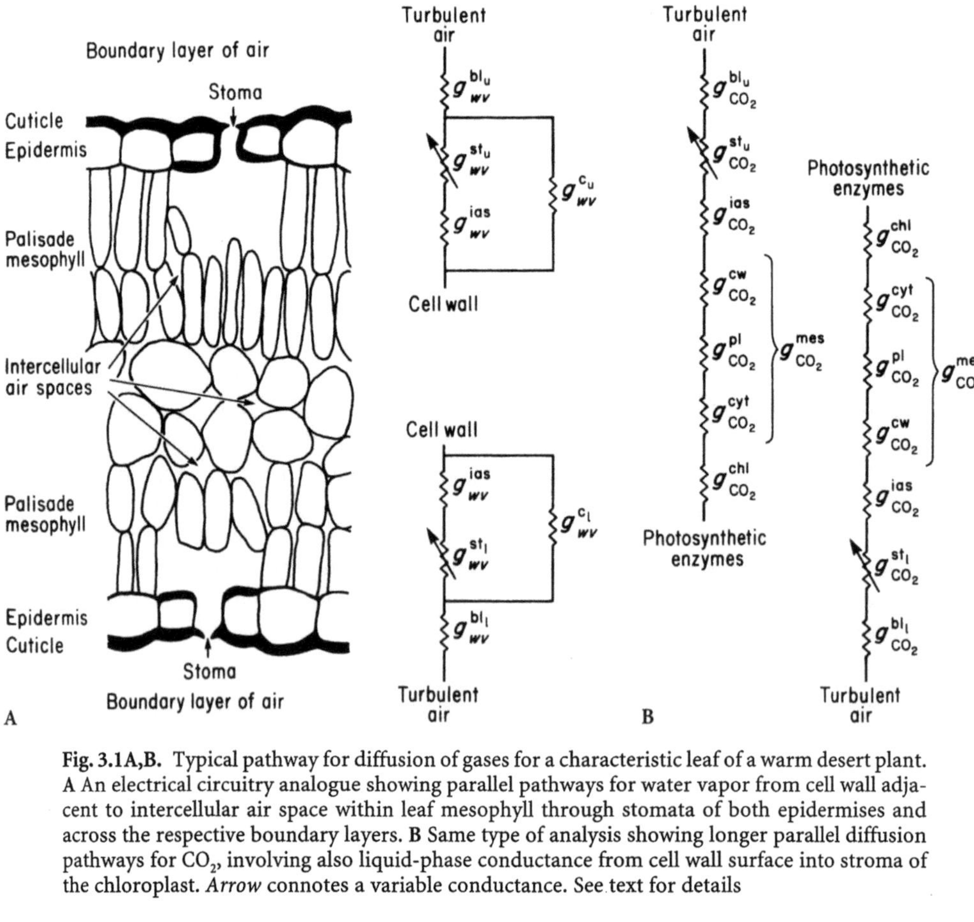

Fig. 3.1A,B. Typical pathway for diffusion of gases for a characteristic leaf of a warm desert plant. **A** An electrical circuitry analogue showing parallel pathways for water vapor from cell wall adjacent to intercellular air space within leaf mesophyll through stomata of both epidermises and across the respective boundary layers. **B** Same type of analysis showing longer parallel diffusion pathways for CO_2, involving also liquid-phase conductance from cell wall surface into stroma of the chloroplast. *Arrow* connotes a variable conductance. See text for details

and the two sides of a leaf also act as parallel conductances and are summed to equal leaf conductance (g_{wv}^{leaf}; Nobel 1991a). Total water vapor conductance (g_{wv}^{total}) involves a boundary layer conductance (g_{wv}^{bl}) in the series for each side of a leaf:

$$g_{wv}^{\text{total}} = \frac{g_{wv}^{\text{leaf}_u} g_{wv}^{\text{bl}_u}}{g_{wv}^{\text{leaf}_u} + g_{wv}^{\text{bl}_u}} + \frac{g_{wv}^{\text{leaf}_l} g_{wv}^{\text{bl}_l}}{g_{wv}^{\text{leaf}_l} + g_{wv}^{\text{bl}_l}} \tag{3.2}$$

Some sensors measure water vapor loss from only one side of a leaf at a time, thus presenting only a portion of transpiration per unit leaf area, whereas cuvettes and sensors that quantify water vapor loss simultaneously from both leaf surfaces (u = adaxial, l = abaxial) are useful to obtain values of g_{wv}^{total}.

3.1.2 Carbon Dioxide

A similar but somewhat longer diffusion pathway series is used to represent CO_2 uptake by a leaf (Fig. 3.1B). As a gas, CO_2 diffuses along the same pathway as water vapor but moves in the opposite direction to cell surfaces of chlorenchyma (mesophyll or stem cortex). There, CO_2 is dissolved in water for a liquid phase as H_2CO_3 or HCO^- to diffuse through cell wall, plasmalemma, and cytosol ($g_{CO_2}^{\text{mes}}$) before moving into the chloroplast ($g_{CO_2}^{\text{chl}}$) for fixation. The expression in series for CO_2 uptake of one leaf or stem surface would be:

$$g_{CO_2}^{\text{leaf}} = \frac{g_{CO_2}^{\text{st}} g_{CO_2}^{\text{ias}} g_{CO_2}^{\text{mes}} g_{CO_2}^{\text{chl}}}{g_{CO_2}^{\text{st}} + g_{CO_2}^{\text{ias}} + g_{CO_2}^{\text{mes}} + g_{CO_2}^{\text{chl}}}, \tag{3.3}$$

and for parallel conductances of two leaf sides with separate boundary layer conductances ($g_{CO_2}^{\text{bl}}$):

$$g_{CO_2}^{\text{total}} = \frac{g_{CO_2}^{\text{leaf}_u} g_{CO_2}^{\text{bl}_u}}{g_{CO_2}^{\text{leaf}_u} + g_{CO_2}^{\text{bl}_u}} + \frac{g_{CO_2}^{\text{leaf}_l} g_{CO_2}^{\text{bl}_l}}{g_{CO_2}^{\text{leaf}_l} + g_{CO_2}^{\text{bl}_l}} \tag{3.4}$$

When stomata are closed, CO_2 uptake would not occur because (1) CO_2 diffusion across the cuticle is extremely slow and (2) CO_2 buildup in tissues from cell metabolism causes a marked decrease in Δc_{CO_2} to the point where an efflux of CO_2 may be expected, when $c_{CO_2}^{\text{cyt}}$ (cy to sol) greatly exceeds ambient CO_2 levels.

3.1.3 Flux Density of Gases

Flux density of a gas, such as water vapor (J_{wv}, mol m^{-2} s^{-1}) or CO_2 (J_{CO_2}), also termed rate of gas diffusion per unit plant surface area, is calculated using a form of Fick's first law:

$$J = g^{\text{total}} \Delta c. \tag{3.5}$$

The highest rates of diffusion occur when drop in concentration is very steep and conductance is high. For water vapor, this condition exists when air is hot and very dry but stomata are wide open for photosynthesis to occur. Little transpiration occurs when conductance is extremely low, e.g., when stomata are closed, or when Δc_{wv} approaches zero, i.e., when stomata are open during morning fog.

3.1.4 Characteristics of Leaf Conductances

3.1.4.1 Leaf Boundary Layer Characteristics

Average thickness of unstirred boundary layer adjacent to each flat leaf surface is approximated as:

$$\delta^{bl} = 4.0\sqrt{\frac{l^{leaf}}{v^{wind}}}, \tag{3.6}$$

where δ^{bl} is boundary layer thickness (m), l^{leaf} is leaf linear dimension in the downwind direction (m), and v^{wind} is ambient wind velocity (Nobel 1991a). Gas molecules diffuse along a concentration gradient and heat is conducted across the boundary layer.

For desert plants with glabrous, microphyllous laminae, the leaf boundary layer thickness in still air (0.1 m s^{-1}) can be up to 3 mm, but during typical windy daytime conditions would be relatively thin, mostly less than 1 mm thick to extremely thin (Table 3.1).

Values of conductance of water vapor (g^{bl}_{wv}) can be estimated using a diffusion equation derived from Fick's first law:

$$g^{bl}_{wv} = \frac{D_{wv}}{\delta^{bl}}, \tag{3.7}$$

Table 3.1. Calculated thickness of boundary layer, δ^{bl} (mm), for flat desert leaves under field conditions, using Eq. (3.5) and specifying leaf linear dimension in the wind direction l^{leaf} (m) and ambient wind velocity, v^{wind} (m s^{-1}). Still air has a wind velocity of approximately 0.1 m s^{-1}

		v^{wind} (m)				
		0.10	0.50	1.00	2.50	5.00
l^{leaf}	0.002	0.57	0.25	0.18	0.11	0.08
(m)	0.005	0.89	0.40	0.28	0.18	0.13
	0.010	1.26	0.57	0.40	0.25	0.18
	0.020	1.79	0.80	0.57	0.36	0.25
	0.050	2.83	1.26	0.89	0.57	0.40
	0.100	4.00	1.79	1.26	0.80	0.57

where D_{wv} is the diffusion coefficient of H_2O at a given temperature and pressure and δ^{bl} is the calculated boundary layer thickness. For water vapor diffusing in midday late-spring desert air at 40 °C, D_{wv} is 2.72×10^{-5} m^2 s^{-1} at 0.1013 MPa air pressure, so that a lamina 5 mm × 5 mm, experiencing a wind speed of 1.00 m s^{-1}, would have a boundary layer conductance of 97 mm s^{-1}. The same leaf under identical conditions would have $g^{bl}_{CO_2}$ ($D_{CO_2} = 1.70 \times 10^{-5}$ m^2 s^{-1}) of 61 mm s^{-1}. Small leaves do not develop low boundary layer conductances, hence the diffusion of water vapor and CO_2 across the boundary layer is significantly more rapid than along the stomatal pores or through the cuticle.

Presence of dense trichomes covering a leaf surface has been explained as a way to decrease boundary layer conductance and thereby reduce leaf transpiration (Oppenheimer 1960; Wooley 1964; McCleary 1968). However, dense hairs on leaves of *Encelia farinosa*, a widespread deciduous shrub of North American deserts, had no physiologically significant effect on the thickness of leaf boundary layer (Ehleringer et al. 1976; Ehleringer 1977, 1983a; Ehleringer and Björkman 1978b; Ehleringer and Mooney 1978).

Effective boundary layer thickness of a cylinder, such as a terete succulent leaf or a relatively smooth photosynthetic stem, is difficult to analyze, but can be estimated by the expression:

$$\delta^{bl} = 5.8\sqrt{\frac{d}{v^{wind}}}, \qquad (3.8)$$

where d (m) is the diameter of the cylinder (Nobel 1991). Hence, a 3-mm-wide green stem of an aphyllous shrub, e.g., *Hammada scoparia* in the Negev or *Senna armata* in California, in a low wind (1 m s^{-1}) would have a δ^{bl} of 0.32 mm, thicker than a flat leaf, but still permitting high g^{bl} of gases.

3.1.4.2 Cuticle and Groundmass Epidermis

Relatively little gas diffusion occurs through groundmass epidermal cells of a leaf or stem when the outer tangential cell walls are impregnated with the water-repellent wax, cutin. Consequently, cuticular transpiration (g^c_{wv}) is extremely low and often requires sophisticated instrumentation to detect. Published values of g^c_{wv} range from 0.01 to 0.4 mm s^{-1}, and xerophytes are known to have some of the lowest values (Nobel 1991a). Lower conductances are often attributed to the presence of a thicker cuticle, but plants of dry habitats also tend to have cuticular waxes that are chemically excellent at reducing transpiration (Martin and Juniper 1970).

Conductances of CO_2 through cuticle are significantly less than those for water vapor conductance, because at the same temperature and atmospheric pressure, the diffusion coefficient of CO_2 (MW 44) is 0.62 that of lighter water vapor (MW 18; Nobel 1991a).

Presence of secreted resins on the cuticle is hypothesized to reduce g^c_{wv}, but resins may also serve to increase reflectance and provide defense against herbivores.

3.1.4.3 Stomata

The chief avenue for gas influx and efflux is through stomata, regulated by guard cells. When open, stomata constitute up to 2% of the leaf area, but values are much lower on young stems. On photosynthetic leaves, stomata may occur on the abaxial surface only (hypostomatic), upper surface only (hyperstomatic or epistomatic), or on both surfaces (amphistomatic) and at different densities and sizes. Stems may lack stomata or develop periderm, both of which sharply limit gas diffusion to internal stem tissues.

The highest values of g^{st}_{wv} are about 20 mm s^{-1}; thus, when stomata are open, cuticular transpiration is a negligible parameter. Stomatal conductance is greatly lowered when hydrostatic pressure in guard cells is decreased slightly and causes partial closure of the pore. Partial stomatal closure, which causes greater interactions between gas molecules and confining cell walls, has the dual consequence of reducing both photosynthetic and transpiration rates.

An application of Fick's first law permits an analysis of significant parameters for the rate of diffusion for any gas species (J_j) through stomata:

$$J_j = \frac{D_j na^{st}}{\delta^{st}} \Delta c_j^{st}, \qquad (3.9)$$

where D_j is the diffusion coefficient of the gas at a specified temperature and pressure, na^{st} is the fraction of leaf surface area for stomatal pores, Δc_j^{st} is the drop in gas concentration through the stomatal pore, and δ^{st} is the depth of the stomatal pore. Theoretically, an epidermis having few large stomatal pores per mm^{-2} could have the same maximum g^{st} with many small stomatal pores.

Standard summaries of xeromorphy have treated the sunken stoma (Fig. 3.2A–D) as an adaptation to reduce transpiration, because this feature is best expressed in plants of dry habitats (Oppenheimer 1960; Fahn and Cutler 1992). Suprastomatal (epistomatal) cavities typically are significantly wider and longer than pores of sunken stomata, hence conductance through cavity apertures would be predicted to be much higher than through associated stomatal apertures, because na^{st} [Eq. (3.9)] is greater. Cavities to the outside of stomatal pores can be responsible for slowing diffusion only when they are very narrow and extremely deep (δ^{st}; Gibson 1983). Stomata sunken to a depth of up to 60 μm have calculated water vapor conductances of 15–80 mm s^{-1}, which certainly exceed stomatal conductances (max. g^{st} is 20 mm s^{-1} for cultivated plants), even when shoot water potential is extremely high. Consequently, when a shoot is under greatest water stress and stomatal conductance is low, suprastomatal conductance would not be expected to reduce transpiration. A very deep (>60 μm), narrow cavity theoretically limits transpiration by lengthening the diffusion pathway, but only when plants are fully hydrated and maintain very high g^{st}. Any reduction of transpiration by this method would also lower CO_2 uptake, although proportionately less because water vapor diffuses along a shorter pathway, making g^{st} more important.

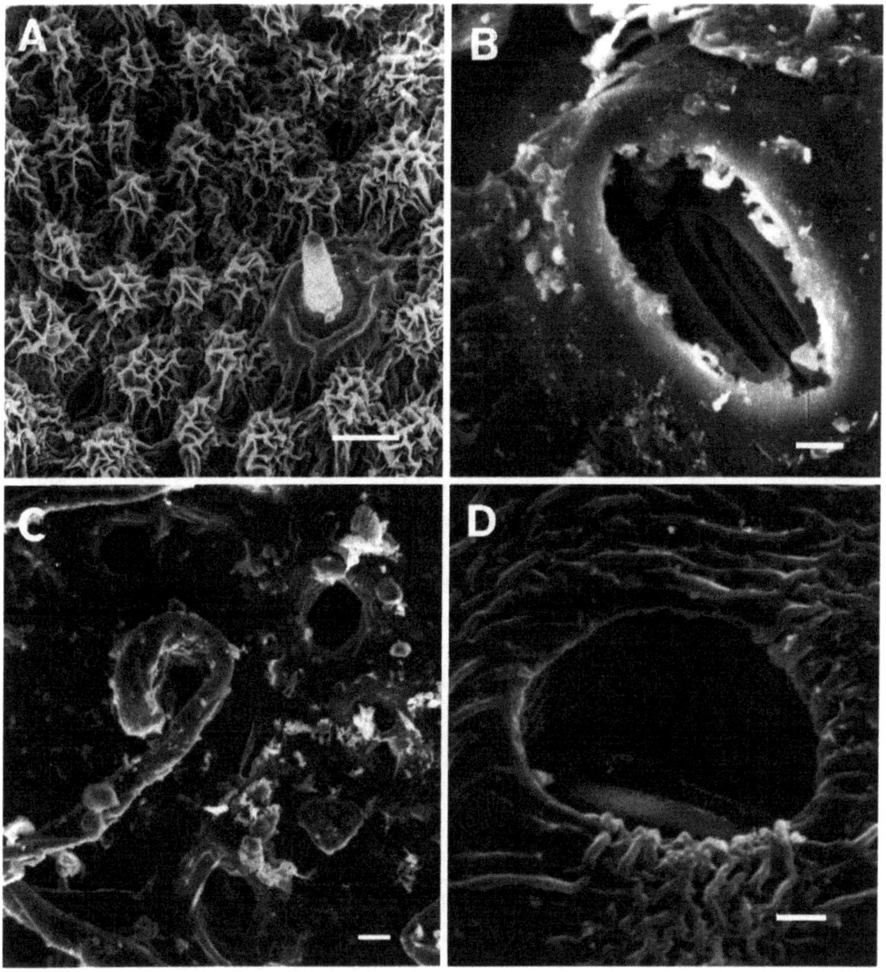

Fig. 3.2A–D. Sunken stomata on adaxial leaf surfaces. **A** *Isomeris arborea*. **B** *Maytenus polyacantha*. **C** *Balanites aegyptiaca*. **D** *Capparis aegyptia*. *Bar* **A** 25 μm; **B–D** 5 μm

If the sunken stoma is not, as traditionally treated, an adaptation to reduce transpiration by lengthening the gas diffusion pathway (e.g., Haberlandt 1914), other hypotheses need to be tested. To increase CO_2 uptake, a deep suprastomatal cavity could be a humidified space, permitting a stoma *to remain open* when, under low relative humidity conditions, it would tend to close (Zeiger 1983; Taiz and Zeiger 1991). Alternatively, it is possible that typical sunken stomata do not significantly modify the gas diffusion pathway directly; instead, the principal adaptation may be the thickened, cutinized epidermis, which thereby affects cuticular conductance, energy budget, or biomechanics of the plant organ, with sunken stomata as a secondary consequence.

3.1.4.4 Intercellular Air Spaces

Between the stomatal pore and internal parenchyma cell walls there is much intercellular air space. Mesophyll parenchyma typically consists of cells fused over only 5–10% of their surface because cells tend to separate during leaf expansion. A substomatal chamber occurs interior to each stoma because no mesophyll cells are attached to guard cells; hence, the width of the chamber exceeds the length or twice the width of a guard cell.

Conductance through intercellular air space (g_j^{ias}) resembles that within an unstirred air layer, and can be calculated:

$$g_j^{\text{ias}} = \frac{D_j}{\delta_j^{ias}}, \tag{3.10}$$

where D_j is the diffusion coefficient of the gas (j) and δ_j^{ias} is the effective length of the diffusion pathway. Because δ_{wv}^{ias} is probably very short, most evaporation coming from cells adjacent to the stomatal pore (Farquhar and Raschke 1978; Appleby and Davies 1983), calculated values for g_{wv}^{ias} are exceedingly high. For CO_2 diffusion in a leaf at 40 °C ($D_{CO_2} = 1.70 \times 10^{-5}\,m^2\,s^{-1}$) with an effective length of 100 μm, $g_{CO_2}^{\text{ias}}$ would be more rapid than even boundary layer, which is the least limiting of the other conductances. Only in leaves where $\delta_{CO_2}^{\text{ias}}$ exceeds 300 μm and at low air temperature would $g_{CO_2}^{\text{ias}}$ be expected to become slow enough for potentially limiting CO_2 uptake, but even under these circumstances $g_{CO_2}^{\text{ias}}$ typically would be greater than $g_{CO_2}^{\text{st}}$. Very thick leaves or stem chlorenchyma are predicted to have rate-limiting $g_{CO_2}^{\text{ias}}$ for photosynthesis.

3.1.4.5 Liquid-Phase Conductance of CO_2

Calculated values of chlorenchyma cell wall conductances for H_2CO_3 and HCO^- ($g_{CO_2}^{cw}$), commencing liquid-phase component of the diffusion pathway, are 100 mm s^{-1} for the cell wall and 30 mm s^{-1} for the plasmalemma (Nobel 1991a). This assumes that cell walls are thin and do not have uncharacteristic wall constituents, such as thick wax or lignin, and considers only cell walls that are exposed to intercellular air space.

If freely exposed internal mesophyll surface is 20 times greater than leaf area ($A^{\text{mes}}/A = 20$), then at 20 °C the estimated CO_2 diffusion from intercellular air space to cytosol ($g_{CO_2}^{cw+\text{pl}}$) would be 7 mm s^{-1}, primarily due to slow diffusion across the plasmalemma (Nobel 1991a). Therefore, plasmalemma conductance, which can be less than stomatal conductance under optimal conditions, probably accounts for some reduction of CO_2 diffusion and is the most important structural parameter within the mesophyll for limiting CO_2 uptake (Nobel et al. 1975; Longstreth et al. 1980; Nobel and Longstreth 1981; Nobel and Walker 1985). When designed for high photosynthetic capacity, a leaf should possess relatively dense chlorenchyma with high A^{mes}/A, therefore great numbers of chloroplasts per unit area. These adaptations are expected wherever the leaf develops under maximal solar irradiation, where PPFD is not the limiting factor for photosynthesis. High A^{mes}/A permits a leaf to be small, minimizing high temperature effects (Chap. 2), and yet have the same total photosynthesis as a much wider, but thinner, leaf.

Chloroplasts typically have a peripheral arrangement along radial walls of palisade parenchyma cells, maximizing the diffusion of CO_2 to stroma once the molecules enter the protoplast. Liquid-phase conductance CO_2 from cytosol into a chloroplast is estimated to be 100 mm s^{-1} (Nobel 1991a).

Comparative studies have shown that chlorophyll content of a leaf is not limiting for photosynthesis but that light-saturated photosynthesis is limited by the concentration of Rubisco (RuBP carboxylase-oxygenase; Björkman 1981). Björkman has concluded that high A^{mes}/A values are important for higher photosynthetic rates primarily because Rubisco is increased with increase in mesophyll volume, and he rejected claims that A^{mes}/A is a physiologically significant parameter because he concluded that internal mesophyll surface has a negligible diffusive resistance of CO_2.

3.1.5 *Factors Limiting Photosynthetic Rates at Saturated Photon Flux Density*

The driving force for CO_2 uptake is created by the consumption of intercellular CO_2 in chlorenchyma cells, and, as just described, the rate of CO_2 uptake per unit leaf area can be strongly affected by: (1) maximal stomatal conductance; (2) total mesophyll cell surface area by which dissolved CO_2 can diffuse to reach the chloroplasts; and (3) density of chloroplasts per unit area, especially as it determines concentration of key photosynthetic enzymes (Rubisco). Such changes can be observed when nondesert plants are grown at an increasingly high PPFD (Chabot and Chabot 1977; Chabot et al. 1979).

3.1.5.1 Relative Effects of Parameters

For photosynthetic tissue, net influx of CO_2 (J_{CO_2}) can be expressed as:

$$J_{CO_2} = \frac{c_{CO_2}^{ta} - c_{CO_2}^{chl}}{r_{CO_2}^{bl} + r_{CO_2}^{leaf} + r_{CO_2}^{mes} + \left(1 + \frac{J_{CO_2}^{r+pr}}{J_{CO_2}}\right) r_{CO_2}^{chl}}, \tag{3.11}$$

where the drop in CO_2 concentration from air ($c_{CO_2}^{ta}$) into the chloroplast ($c_{CO_2}^{chl}$) is divided by the resistances of the gas phase, i.e., boundary layer ($r_{CO_2}^{bl}$) through stomata and intercellular air space ($r_{CO_2}^{leaf}$), plus liquid-phase mesophyll resistance ($r_{CO_2}^{mes}$), plus chloroplast resistance ($r_{CO_2}^{chl}$) but then reduced by losses of carbon via cellular respiration and photorespiration ($J_{CO_2}^{r+pr}$; Nobel 1991a). Analyses of these parameters reveal where factors probably set limits on the photosynthetic capacity of an assimilatory organ.

Suppose that a leaf with A^{mes}/A of 20 has a fairly typical c_{CO_2}, $r^{bl}_{CO_2}$ is 60 s m^{-1}, $r^{leaf}_{CO_2}$ is 290 s m^{-1} and with only abaxial stomata, $r^{mes}_{CO_2}$ is 140 s m^{-1}, $r^{chl}_{CO_2}$ is 100 s m^{-1}, and respiratory losses are 30%:

$$J_{CO_2} = \frac{6.8 \times 10^{-5}\ \text{mol m}^{-3}}{(1.00)\ (350\ \text{s m}^{-1}) + 140\ \text{s m}^{-1} + (1.0+0.3)(100\ \text{s m}^{-1})}$$

$$= 11.0\ \mu\text{mol m}^{-2}\ \text{s}^{-1}\ \text{(hypostomatic or epistomatic)}$$

By having equal CO_2 conductance from each side (amphistomaty) of that leaf:

$$J_{CO_2} = \frac{6.8 \times 10^{-5}\ \text{mol m}^{-3}}{(0.50)\ (350\ \text{s m}^{-1}) + 140\ \text{s m}^{-1} + (1.3)\ (100\ \text{s m}^{-1})}$$

$$= 15.3\ \mu\text{mol m}^{-2}\ \text{s}^{-1}\ \text{(amphistomatic, 0.50/0.50).}$$

In this example, boundary layer resistance is cut in half, thereby resulting in a higher uptake of CO_2 for the amphistomatic model (Mott et al. 1982), here by nearly 40%. When gas-phase conductance is less on one leaf surface than the other, e.g., 0.30/0.70, the resistance parameter is proportional to the greater value.

Using the amphistomatic leaf, when A^{mes}/A is doubled to 40, thereby theoretically decreasing r^{mes} to 70 s m^{-1}, then:

$$J_{CO_2} = 18.0\ \mu\text{mol m}^{-2}\ \text{s}^{-1}\ \text{(amphistomatic and high } A^{mes}/A\text{)}$$

Whenever A^{mes}/A increase results in significantly thicker mesophyll, then the gas-phase resistance parameter would also increase slightly. Nonetheless, for a light-saturated leaf, structural modifications, such as how stomata are distributed and the amount of internal surface of chlorenchyma, appear to have a profound effect on photosynthetic capacity and net CO_2 uptake per unit leaf area without any changes in guard cell or chloroplast physiology (Fig. 3.3).

3.1.5.2 C_4 Physiology and Elimination of Photorespiration

For plants with C_3 photosynthesis, more than 25% of fixed carbon is simultaneously lost through cellular respiration, mostly photorespiration (Taiz and Zeiger 1991). Photorespiration occurs because Rubisco, the critical first enzyme of C_3 photosynthetic CO_2 fixation dark reaction, also functions as an oxygenase. Within the chloroplast, Rubisco catalyzes RuBP (ribulose 1,5-bisphosphate) and O_2 to combine and form phosphogylcerate, which is exported for subsequent metabolism to mitochondria and peroxisomes.

Extensive reviews have described a totally different approach used by C_4 plants to increase net CO_2 fixation rates by eliminating photorespiration

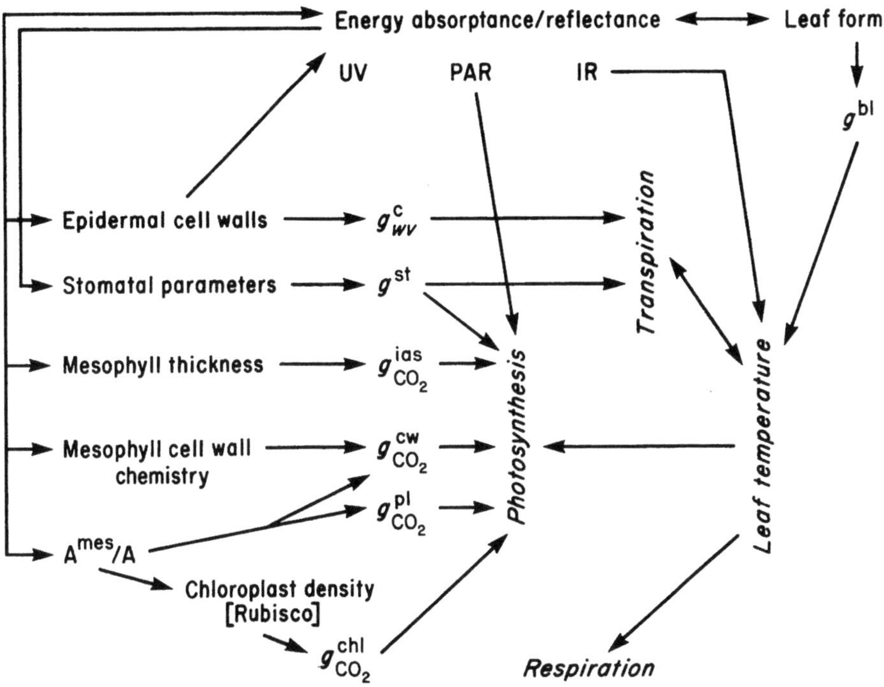

Fig. 3.3. Model summarizing the characteristic effects that anatomical features of leaves have on conductances of gases, which in turn determine rates of photosynthesis and transpiration. Amount and quality of absorbed energy during leaf differentiation modify each of the anatomical parameters, and leaf form is, in part, determined by light levels. Leaf form (size and shape, orientation, and surface features) affects physiological rates by influencing amount and quality of absorbed energy and boundary layer conductances (Chap. 2). Additional factors that could be added to such a model are effects of shoot water status, salinity, or available nutrients. Boundary layer can also slow diffusion of gases as it becomes thicker, but must be very thick to have a substantial physiological effect on the conductance pathway. Changes in photosynthesis, respiration (especially photorespiration), and transpiration will strongly impact WUE

(Lorimer 1981; Ogren 1984; Taiz and Zeiger 1991). Some C_4 plants have among the highest recorded photosynthetic rates, and several examples are desert plants (Nobel 1980d; Ehleringer 1983b, 1994), but it should be understood that C_4 photosynthesis is not a desert adaptation per se, but evolved in other tropical and subtropical ecosystems and simply characterized numerous successful immigrants to the desert environment. Nonetheless, C_4 photosynthesis is an appropriate desert adaptation because these plants grow extremely well in full-sun habitats that are both hot and dry.

Briefly stated, C_4 photosynthesis utilizes a spatial separation within the leaf of CO_2 fixation and reduction to a carbohydrate, and each phase uses a different enzyme. Fixation of CO_2 occurs in chloroplast-containing mesophyll cells (PCA, primary carbon assimilation tissue), where the 3-carbon phosphoenolpyruvate (PEP) is combined with CO_2 to form oxaloacetate, in the presence of PEP carboxylase, and subsequently either malate or aspartate, 4-carbon acids. This process is very efficient and does not lead to loss of carbon via oxygenation. Either malate or aspartate is transported to enlarged cells of the bundle sheath, called Kranz cells (Kranz anatomy or PCR, photosynthetic carbon reduction tissue; Fig. 3.4), presumably mostly via plasmodesmata. Within these special bundle sheath cells, having extremely large chloroplasts but lacking typical granal structure, C_4 acids are decarboxylated in cytoplasm. This generates a relatively high concentration of CO_2 in bundle sheath cells, which enables Rubisco to function largely as a carboxylase, while CO_2 is incorporated within the agranal chloroplasts into sugars using normal biochemical steps of C_3 dark reaction photosynthesis. Photorespiration is thus essentially eliminated within bundle sheath cells, and what little CO_2 is released by bundle sheath metabolism is immediately reincorporated into sugars. The 3-carbon acids produced in the bundle sheath, when malate or aspartate are decarboxylated, are transported back to mesophyll cells for regenerating PEP. Sugars made in bundle sheath cells can be exported efficiently via nearby sieve tubes.

Regarding the previous discussion of Eq. (3.11), it can be shown that improved uptake of CO_2 affects the $r^{chl}_{CO_2}$ term in the equation. Essentially all C_4 leaves, and especially those in high light environments, are amphistomatic, so there is high CO_2 uptake per degree of opening of stomata. However, C_4 leaves are comparatively thin and have low values of A^{mes}/A (Pearcy et al. 1981), because C_4 Kranz anatomy usually has CO_2-fixing mesophyll cells that are radially arranged around the bundle sheath, and requires $g^{ias}_{CO_2}$ to be high with its short effective length for CO_2 diffusion.

The C_4 pathway is especially adaptive for tropical and subtropical desert environments because photosystems generally have an extremely high light

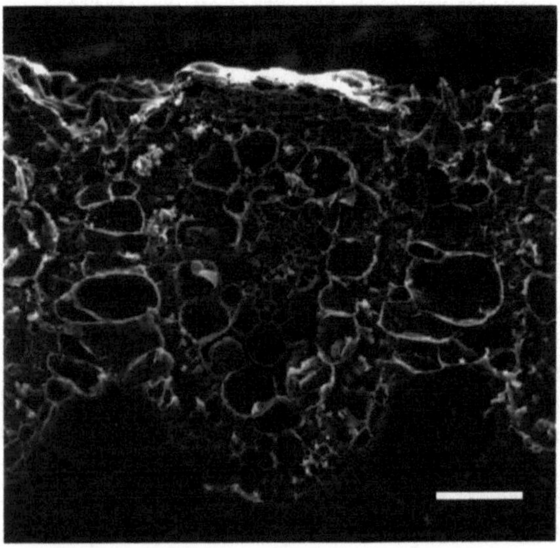

Fig. 3.4. Kranz anatomy of the C_4 leaf of *Pleuraphis rigida*, which has enlarged bundle sheath cells, enclosing a vascular bundle and itself surrounded by mesophyll parenchyma. *Bar* 50 µm

saturation point and temperature optima for photosynthesis tend to be comparatively high (Taiz and Zeiger 1991), hence they can obtain higher yields under summer conditions if water is available.

3.1.6 Water-Use Efficiency

Water-use efficiency (WUE), a ratio of mass or moles of CO_2 fixed per unit of water vapor transpired, or its reciprocal, transpiration ratio, is a useful measure of a plant's response to onset of drought, and a way to assess importance of structural adaptations to gas diffusion, especially to diagnose any temporal, spatial, or evolutionary trends of water conservation. Many factors that influence WUE are physiological properties of guard cells, which are not part of the current discussion, e.g., effects of CO_2 concentrations, light quantity, water vapor concentration and relative humidity, and abscisic acid on stomatal aperture (Zeiger 1983; Taiz and Zeiger 1991), and physiologists have developed models to test optimal behavior of stomata to minimize water loss while maximizing carbon gain (Cowan 1977, 1981; Cowan and Farquhar 1977).

As a general rule of thumb, 24-h WUE values of actively growing C_3 plants are 0.001 to 0.003 g g^{-1} (=4 to 12 mmol CO_2 mol^{-1} H_2O), those of C_4 plants 0.002 to 0.005 g g^{-1}, and of CAM plants 0.01 to 0.06 g g^{-1} (Szarek and Ting 1975; Nobel 1988, 1991b). By lowering temperature of the plant organ, J_{wv} is strongly decreased because c_{wv}^{ias} is greatly reduced, therefore Δc_{wv} is decreased [Eq. (3.5)], this without affecting CO_2 diffusion, which is less influenced by small temperature changes. CAM plants have very high WUE because stomata open at night, when air is relatively cooler, and Δc_{wv} is significantly smaller than during daylight (Gibson and Nobel 1986).

Changes in structural parameters can yield a substantial increase in WUE whenever J_{wv} is reduced at a higher rate than J_{CO_2}. Although both gases diffuse through the same stomatal pores and across the same boundary layer, the pathway for water vapor is much shorter, and therefore any effect on stomatal conductance or boundary layer thickness has a greater fractional influence on J_{wv} than J_{CO_2}. Hypothetically, narrow leaf design should yield a somewhat lower WUE than a wide one, given that the boundary layer is much thinner (Table 3.1). Conversely, a leaf that has partially closed stomata increases stomatal resistance to water vapor more than for CO_2; this is cited as a reason why WUE of C_4 leaves is typically higher than for C_3 leaves, and probably why leaves under water stress with partial stomatal opening experience some improvement in WUE (Nobel 1991a).

The r_{wv} of one stomatal pore is a function of pore diameter, so that, on average, the widest pore is predicted to have a lower WUE than a very narrow one, in the same way that a fully opened pore has a lower WUE than one partially closed. On an epidermal surface, a field of fully open narrow stomata should have greater WUE than a field of fully open and fewer wide stomata having the identical J_{CO_2}, because transpiration is generally linearly related to vapor pressure deficit.

3.2 Physiological Anatomy of Nonsucculent Leaves of Desert Plants

3.2.1 Deciduous and Evergreen Shrubs

Nonsucculent woody C_3 dicotyledons of full-sun desert habitats share many common anatomical features of their leaves. Upper canopy leaves have a basic design to maximize carbon uptake during those brief times during the year when adequate soil water is available for stomatal opening to occur.

3.2.1.1 Groundmass Epidermis and Trichomes

Leaves of lowland desert shrubs and subshrubs have uniseriate adaxial and abaxial epidermis. A reported exception is the Saharan shrub *Ziziphus lotus*, with biseriate adaxial epidermis (Stocker 1974a), and biseriate adaxial epidermis occurs only along leaf margins in the caudiciform *Adenium boehmianum*. There is no evidence that multiple epidermis is a desert adaptation to reduce transpiration, as suggested earlier by Sabnis (1919–1921). Effects of multiple epidermis on leaf water relations have not been quantified.

For full-sun leaves, groundmass cells of both epidermises tend to be equal in height or slightly larger on the adaxial side (Sabnis 1919–1921). For 51 common species of North American woody dicotyledons (26 families), epidermal thickness rarely exceeded 30 μm, and mean and median adaxial/abaxial heights were 1.21 and 1.15, respectively (Fig. 3.5). In paradermal section, mean epidermal thickness mostly was 15–25 μm but was thinnest in the evergreen *Simmondsia chinensis* (8 μm) and thickest in the drought-deciduous *Lepidium fremontii* (30 μm); cell length rarely exceeded twice the width. Adaxial epidermis was about three times thicker in *Celtis pallida*, a condition that also appears in leaves of shrubby *Grewia* and *Ziziphus* of Old World deserts, where large epidermal cells have been interpreted as having water storage function (Sabnis 1919–1921).

Authors traditionally have concluded that desert plants, and xerophytes in general, have small epidermal cells, and Maximov (1929, 1931) observed that leaves that differentiated under water stress showed decrease in cell size. Additional studies of sun and shade leaves within species consistently demonstrated that sun leaves are more xeromorphic (Shields 1950; Napp-Zinn 1973, 1974; Fahn and Cutler 1992).

Broad comparative data are not available to analyze whether cell volumes of desert-scrub plants are significantly smaller in comparison with those of other vegetation types, although the ability to survive drought has been shown

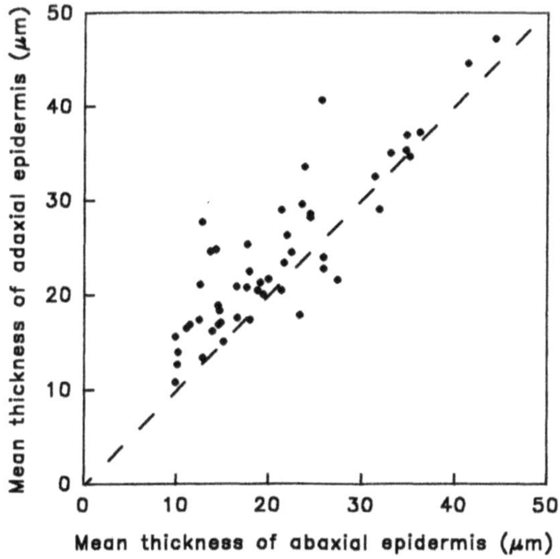

Fig. 3.5. Ratio of adaxial/abaxial epidermal thickness for 51 species of nonsucculent woody perennials of the Sonoran Desert. Nearly all species had ratios >1.0 (---), and the two parameters are highly correlated

to be correlated with small cell volumes (Iljin 1957). Nonetheless, a small cell is not better than a large one at maintaining cell turgor because cell size per se cannot have a direct effect (Cutler et al. 1977). A large cell can lose more water than a smaller cell without changing turgor, although in a study on mesomorphic leaves of *Cyclamen persicum*, cells with smallest volume were most resistant to plasmolysis (Oertli 1986). Smaller cells are more resistant to collapse (cytorrhysis) under water stress, if cell wall thickness is the same, and if smaller cells have slightly thicker walls, the effect would be greater (Cutler et al. 1977).

From the same 51 North American species studied above, 78% of the species had straight or slightly curved anticlinal walls and only 11 species had strongly undulating or sinuous walls. Those with undulating anticlinal walls mostly occurred in Asteraceae, Lamiaceae, and Mimosoideae of Fabaceae, and all these cases had extremely thin anticlinal walls. In deserts of the Middle East and northern Africa, undulations characterize leaf epidermis in species of *Teucrium*, *Artemisia*, and *Chrozophora*, and in the Thar, Fabaceae and Capparaceae often show this feature (Sabnis 1919–1921). Straight anticlinal walls and early onset of wall thickening are most likely correlated. In desert shrubs, undulating walls can frequently be observed in juvenile or seedling leaves, and probably are more commonly seen on abaxial epidermis.

Many xerophytes have thick cuticles and thick outer periclinal epidermal cell walls (Fahn and Cutler 1992). Examples of thick primary cuticle can be found among deciduous species, e.g., strongly papillose leaves of *Becium burchellianum* (Fig. 2.16C), and yet on most nonsucculent desert leaves, primary cuticle is not excessively thickened. Instead, this cuticle tends to be thin and only rarely exceeds 2 µm. More commonly, the outer periclinal cell wall is heavily cutinized, and thus the quality of the wax probably yields characteristic low g_{wv} when stomates are tightly closed (Kurtz 1958). Relatively thick primary cuticle commonly occurs among drought-deciduous Fabaceae, especially *Acacia*, *Calliandra*, *Olneya*, and *Prosopis* (Bleckmann and Hull 1975; Mooney et al. 1977b), and, in desert representatives of *Mortonia* and *Condalia*, crystal sand, characteristic of selected families, is deposited in the cellulosic walls (Metcalfe and Chalk 1950).

Examples of thick secondary cuticle (cuticularized cell wall >5 μm) are found among evergreen shrubs (Fig. 3.6), e.g., *Mortonia scabrella*, *Simmondsia chinensis*, *Leucophyllum frutescens*, *Capparis aegyptia* (Volkens 1887), *Viscainoa geniculata*, *Ericameria linearifolia*, *Zygophyllum prismatocarpum*, *Maytenus polyacantha*, and *Chrysothamnus paniculatus*, but thick outer cell walls also may be found in ephemeral leaves, e.g., those of the aphyllous shrubs

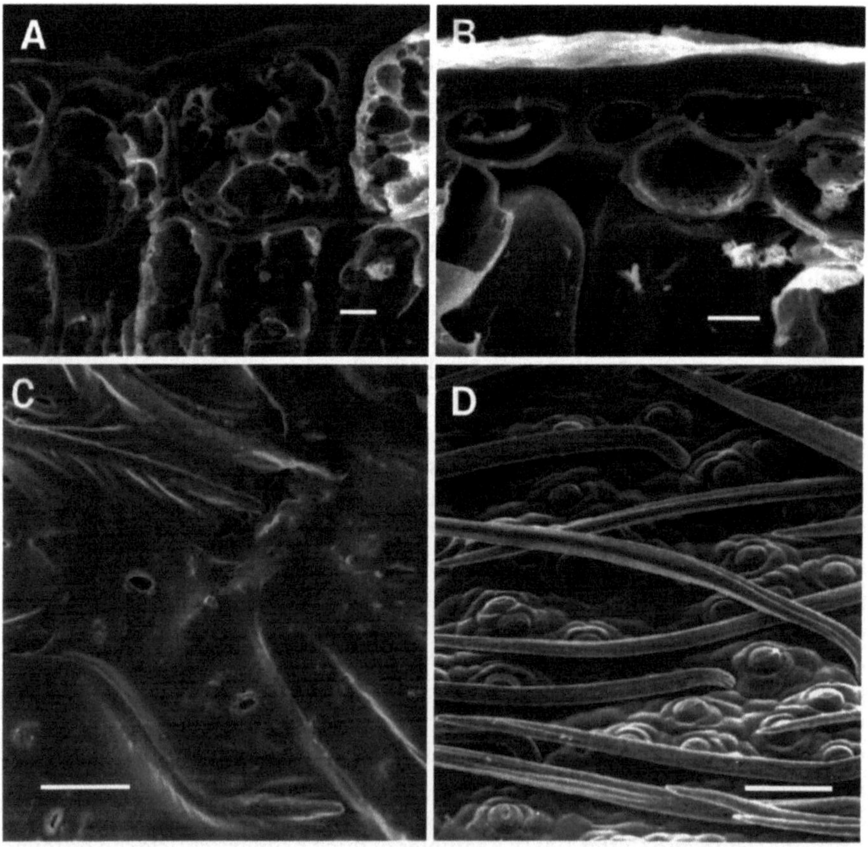

Fig. 3.6A–D. **A** *Larrea tridentata* with thickened, periclinal epidermal cell walls and resin within epidermal cells. **B** *Maytenus polyacantha*; this evergreen leaf has thick epidermal cell walls and a uniseriate hypodermis. **C,D** *L. tridentata*; **C** adaxial epidermis has trichomes entrapped in its resin-coated surface; **D** same as **C** with resin removed, revealing slightly raised guard cells. Judging from the high stomatal densities in **D**, numerous stomata in **C** appear to be covered by the resin. *Bar* **A,B** 5 μm; **C,D** 50 μm

Castela polyandra and *Glossopetalon nevadensis*, or drought-deciduous leaves, e.g., *Acamptopappus sphaerocephalus* and *Ipomopsis gloriosus*, indicating that leaf phenology is not tightly correlated with heavy wax deposition. In fact, many common desert chamaephytes, e.g., species of *Ambrosia, Artemisia, Chrozophora*, and *Encelia*, have combined cuticle and outer wall thickness that rarely exceeds 2 µm. On a given leaf, cells with the thickest outer tangential cell walls occur along leaf margins.

Anticlinal and inner periclinal cell walls of leaf epidermis may also be thickened, probably slowing cuticular transpiration (Sabnis 1919–1921). Thickened cell walls are especially pronounced in evergreen leaves of *Larrea tridentata* (Fig. 3.6A; Thomson et al. 1979), *Z. prismatocarpum*, *M. polyacantha* (Fig. 3.6B), and *Leucophyllum frutescens*, but inner walls 2–3 µm thick can also be observed in deciduous leaves of *Chilopsis linearis, Cercidium microphyllum*, and *Lepidium fremontii*, and in many species along leaf margins. Some cell walls of legumes may be mucilaginous and thereby have higher matrix potential (Volkens 1887; Sabnis 1919–1921). No study has determined to what degree inner periclinal walls are cutinized, but the presence of cutin or suberin is probably a crucial adaptation for resisting water stress.

Wall thickness can influence turgor pressure, in that thin-walled cells can lose water as much as thick-walled cells without changing turgor, so that as leaf water potential decreases, turgor can be maintained more easily in a leaf with mostly thin-walled cells (Cutler et al. 1975). A thick-walled cell will lose turgor more rapidly because it has a higher volumetric elastic modulus, i.e., it is more rigid [Eq. (2.28) in Nobel 1991a]. Nonetheless, known adaptations for maintaining turgor are mainly those of osmotic adjustment via biochemical mechanisms (Turner and Jones 1980; Nilsen et al. 1981).

Basal cells of trichomes may be cutinized to reduce evaporation (Fahn 1986; Fahn and Cutler 1992). In *Larrea tridentata*, thick trichome walls contain a suberin layer (Thomson et al. 1979).

A small number of woody perennials have leaves that are coated with resin. Famous examples are the species of *Larrea, Acacia neovernicosa, Ambrosia deltoidea*, species of *Haplopappus* (*Ericameria*) and *Chrysothamnus, Dodonaea microgyna, Baccharis sarothroides, Hymenoclea salsola, Rhus burchelli* and *Eriocephalus ericoides* (Jordaan and Kruger 1992), and *Flourensia cernua*, and surface resins are common among phyllodineous *Acacia* of central Australia (Boughton 1990)

In *Larrea tridentata*, groundmass epidermal cells and trichomes produce resin precursors (Fig. 3.6A), which are extruded to the surface as liquid and there harden, cementing the trichomes (Fig. 3.6C–D; Thomson et al. 1979). The principal phenolic constituent is nordihydroguaiaretic acid (Mabry et al. 1977), and recent studies have demonstrated surface resin on these leaves of *Larrea* reduces transpiration (Meinzer et al. 1990), even though the cuticle is thin and is only one-sixth the amount of outer wall material (Hull et al. 1971). Glandular trichomes produce terpenoid-rich surface resins in numerous Asteraceae, but effects of these resins on the gas diffusion pathway have not been quantified.

Guard cells of these plants tend to be raised slightly or have projecting outer ledges (Figs. 3.6D), and thereby prevent most surface resin from occluding stomata (Dell and McCoomb 1978).

3.2.1.2 Stomata

Since Volkens (1887), many authors have observed that nonsucculent leaves of woody desert plants are amphistomatic, each surface having very similar densities of abundant, small stomata (Table 3.2). Moreover, in leaves from Old and New World deserts, authors discovered that, for many species, guard cells on leaves are not sunken or become sunken only a few micrometers. Given that botanists expected to find leaves with obvious features to limit transpiration, i.e., few and hidden stomata, thick cuticle and cell walls, and abundant

Table 3.2. Species occurring in warm deserts having either amphistomatic or hypostomatic nonsucculent to fleshy (*) leaves, grouped by mean thickness of lamina (excluding trichomes and midvein measurements). Leaves were sampled from fully expanded leaves on rapidly growing, upper canopy, full-sun plants, i.e., from environments in which leaf morphogenesis would have occurred under conditons of high total daily PPFD and without obvious effects of water stress. Data for most African species and from the Negev were obtained from published reports (Volkens 1887; Cannon 1924; Stocker 1970, 1971, 1972, 1974a; Evenari et al. 1971). Excluded from the list are Arecaceae, Poaceae, rosette monocotyledons, and aquatic plants of desert springs

Hypostomatic

Thickness of lamina <200 μm

Berberis fremontii
B. haematocarpus
Canotia holacantha
Frankenia pulverulenta
Janusia gracilis
Morus microphylla
Pachycormus discolor
Rhus microphylla
R. tripartita
Sageretia wrightii
Thymelaea hirsuta

Thickness of lamina 200–400 μm

Adenium boehmianum
Berberis trifoliolata
Celtis pallida
Coleogyne ramosissima
Condalia correllii
C. fasciculata (stomatal crypts)
Dendrosicyos socotrana
Dorstenia gigas
Ficus brandegeei
Forestiera neomexicana
Frankenia quadrifolia
F. uncinata
Grewia robusta
Ozoroa dispar
Sophora secundiflora
Ziziphus lotus

Thickness of lamina 401–600 μm

Atamisquea emarginata
Koeberlinia spinosa
Purshia tridentata var. *glandulosa*
Rhus lenti

Amphistomatic

Thickness of lamina <200 μm

Acacia greggii
A. karroo
Ambrosia ambrosioides
A. cordifolia
A. deltoidea
Artemisia herba-alba
Atriplex hymenelytra
Astragalus spinosus
Bursera hindsiana
Calliandra eriophylla
Cercidium floridum
Chamaesyce albomarginata
Eriogonum wrightii
Filago depressa

Table 3.2 (*contd.*)

Gourliea chilensis
Heliotropium arbainense
Ifloga spicata
Lepidium lasiocarpum
Mimosa biuncifera
Pandorea doratoxylon
Phagnalon barbeyana
Pluchea sericea
Plumbago aphylla
Poliomintha incana
Prosopis kuntzei
Prunus fasciculata
Psathyrotes ramosissima
Reichardia tingitana
Roemeria hybrida ssp. *dodecandra*
Solanum lasiophyllum
Stachys aegyptiaca
Tecoma stans var. angustata
Teucrium leucocladium
T. pilosum
Tidestromia lanuginosa
Trixis californica
Viguiera deltoidea var. *parishii*

Thickness of lamina 200–400 µm

Abutilon muticum
A. pringlei
Acacia angustissima
A. constricta
A. neovernicosa
A. raddiana
A. senegal
A. tortilis
Acamptopappus sphaerocephalus
Aloysia wrightii
Argemone corymbosa
Artemisia tridentata ssp. *parishii*
Asclepias erosa
A. filiformis
Atriplex barkleyana
A. canescens
A. halimus
A. vesicaria
Baccharis emoryi
B. sarothroides
Bassia muricata
Bebbia juncea var. *aspera*
Becium burchellianum
Brickellia arguta
B. desertorum
Bulnesia retama
Bursera microphylla
Caesalpinia virgata
Camissonia boothii
C. brevipes
Caulanthus inflatus
Caylusea hexagyna
Cercidium microphyllum
C. praecox
Chilopsis linearis
Chloracantha spinosa
Chrozophora brocchiana
C. obliqua
Cleome droserifolia
Clianthus formosus
Cocculus pendulus
Commicarpus scandens
Convolvulus lanatus
Corchorus antichorus
Crossosoma bigelovii
Croton californica ssp. *mohavensis*
Cryptantha micrantha
Cucurbita digitata
C. palmata
Dalea bicolor var. *orcuttiana*
Datura wrightii
Dicoria canescens
Dodonaea microgyna
D. viscosa var. *angustifolia*
Encelia actonii
E. farinosa
E. frutescens
Eriocephalus africanus
Eriogonum fasciculatum var. *polifolium*
E. inflatum
Eschscholtzia glyptosperma
E. minutiflora
Euphorbia lactiflua
E. tomentulosa
Fagonia arabica
F. laevis
F. pachyacantha
Flourensia cernua
Fouquieria burragei
F. diguetii
F. shrevei
F. splendens
Glossopetalon nevadensis
Gomphocarpus sinaicus
Guaiacum coulteri
Gutierrezia microcephala
G. sarothrae
Haplophyllum tuberculatum
Helianthemum kahiricum
Hymenoclea monogyra
H. salsola
Hyptis emoryi
Jatropha cardiophylla
J. cinerea
J. macrorhiza
Justicia californica
Krameria grayi

Table 3.2 (*contd.*)

K. erecta
Larrea tridentata
Leucophyllum frutescens
Lomatium mohavensis
Lupinus concinnus
Lycium brevipes
Malacothrix glabrata
Matthiola livida
Menodora scabra
M. spinescens
Mimulus bigelovii
Moltkiopsis ciliata
Monoptilon bellioides
Nemacladus rubescens
Neurada procumbens
Olneya tesota
Oxytheca perfoliata
(basal leaves and photosynthetic bracts)
Pachypodium namaquanum
Parthenium incanum
Pectis papposa
Pectocarya recurvata
Penstemon palmeri
Phacelia corrugata
P. distans
Plantago ovata
Polygonum equisetiforme
Populus fremontii
P. wislizenii
Prosopis velutina
Psorothamnus polyadenia
Reseda arabica
R. muricata
Salazaria mexicana
Salix goddingii
Salvadora persica
Salvia apiana
S. aegyptiaca
S. vaseyi
Scrophularia deserti
Senna italica
S. nemophila
Sisyndite spartea
Sphaeralcea ambigua
Stanleya pinnata
Stillingia spinulosa
Tetradymia glabrata
Tetragonia maritima
Thamnosma montana
Tidestromia oblongifolia
Tiquilia plicata
Tribulus longipetalus
Trigonella stellata
Viguiera reticulata
Viscainoa geniculata
Xylorhiza tortifolia
Zinnia acerosa
Z. pumila
Ziziphus obtusifolius var. *canescens*
Z. spina-christi

Thickness of lamina 401–600 μm

Achyronychia cooperi
Allium desertorum
Ambrosia dumosa
Anulocaulis annulata
Asclepias subulata
Balanites aegyptiaca
Boscia senegalensis
Carduncellus eriocephalus
Castela polyandra
Chorizanthe rigida (bracts)
Citrullus colocynthis
Crotalaria aegyptiaca
Dipcadi erythraeum
Diplotaxis acris
Echiochilon fruticosum
Eremophila glabra
Ericameria laricifolia
Eriogonum brachypoda
Fouquieria columnaris
Grayia spinosa
Gymnarrhena micrantha
Harfordia macroptera
Hesperocallis undulatus
Isocoma acradenia
Isomeris arborea
*Lepidium flavum**
*L. fremontii**
Lotus humistratus
L. tomentellus
*Lycium shawii**
Mirabilis bigelovii
Mortonia sempervirens
*Nitraria retusa**
*Nolana sedifolia**
Oligomeris linifolia
Peganum harmala
Petalonyx thurberi
Polycarpaea spicata
Prosopis glandulosa ssp. *torreyana*
Psorothamnus emoryi
P. spinosus
Pteranthus dichotomus
Reaumuria hirtella
Rosenia humilus
Rumex hymenosepalus
Salvia dorrii
Senna armata
S. purpusii

Table 3.2 (*contd.*)

Simmondsia chinensis	*Ipomopsis gloriosa*
Stephanomeria pauciflora	*Lupinus microcarpus* var. *horizontalis*
Stillingia paucidentata	*Lycium andersonii**
	*Maireana sedifolia**
Thickness of lamina > 600 μm	*Maytenus polyacantha*
	*Nitraria schoberi**
Abronia villosa	*Nolana mollis**
Allium rothii	*N. sedifolia**
Berkheya fruticosa	*Ochradenus baccatus**
Calotropis procera	*Pentzia incana*
Capparis aegyptia	*Phaulothamnus spinescens*
C. spinosa	*Suaeda moquinii**
Chorizanthe rigida (basal leaves and bracts)	*Wahlenbergia* sp.
Chrysothamnus paniculatus	*Welwitschia mirabilis*
Cleome arabica	*Zilla spinosa* (winter leaves)
Ericameria linearifolia	*Zygophyllum album**
Farsetia aegyptiaca	*Z. dumosum**
Gymnocarpos decandrum	*Z. prismatocarpum*
Gypothamnium pinifolium	*Z. simplex**
Heliotropium pycnophyllum	*Z. stapffii**
Iphiona mucronata	*Z. waterlotii**

trichomes, authors have often apologized when desert leaves lacked these characters, or often labeled these examples as "mesomorphic" (Cannon 1924; Runyon 1934; Humphrey 1935; Evenari 1938b). For *Bauhinia marlothii* of the Namib, having amphistomatic leaves with relatively thin outer epidermal walls, small and superficially placed stomata, and without trichomes, Cannon stated that leaf anatomy "did not suggest the extreme aridity of the habitat".

More than 90% of nonsucculent woody perennials in hot inland deserts, and in particular the codominants in each community, have amphistomatic leaves (Table 3.2).

Characteristic stomatal densities of widespread species are 100 or more stomata per mm^{-2} per surface, and some exceed 400 mm^{-2} (Waisel 1963; Sundberg 1986), and densities are greater on desert than mesophytic ecotypes (Waisel 1963). Stomatal densities of the two surfaces seldom differ by more than 10% for leaves that developed under full-sun conditions, regardless of whether leaves are erectophilic or planophilic, although there tend to be more stomata on the abaxial surface (Sabnis 1919–1921).

Hypostomaty (Table 3.2) has been observed in deserticolous species of Anacardiaceae, *Pachycormus discolor* (Gibson 1981), *Rhus tripartita* (Stocker 1974a), *R. microphylla*, and *Ozoroa dispar*, but apparently leaves of *R. mucronata* and *R. burchelli* are amphistomatic (Zemke 1939; Jordaan and Kruger 1992). Adaxial stomata are absent in *Thymelaea hirsuta* (Stocker 1974a), which evolved from a genus typical of maquis vegetation, and *Coleogyne ramossissima*, which is most common in semiarid upland sites. Among Rhamnaceae, several species of *Ziziphus* (Sabnis 1919–1921; Stocker 1974a), *Condalia fasciculata*, *C. correllii*, and *Sageretia wrightii* are hypostomatic, whereas *Z. spina-cristi* and *Z. obtusifolia* are amphistomatic. Sabnis observed that hypostomaty is most common for leaves with extremely large adaxial epidermal cells. In nearly all cases of hypostomaty on full-sun desert leaves, adaxial epidermis is at least twice the thickness of abaxial epidermis.

Although adaxial/abaxial stomatal density ratios are approximately unity for typical full-sun leaves of desert species (Wood 1934), ratios may be less than 0.5 for leaves that develop in partial shade of the canopy, and on full-shade leaves of *Ambrosia cordifolia* adaxial stomata may be absent (Mott et al. 1982; Mott and Michaelson 1991). This facultative response supports a hypothesis that amphistomaty is an adaptive strategy for leaves experiencing and efficiently utilizing high levels of PPFD. For xerophytes with hypostomatic drought leaves, researchers have demonstrated amphistomatic wet-season leaves in *Phlomis fruticosa* (Christodoulakis 1989). Several species reported from cultivated, probably shaded plants as having hypostomatic leaves (Sundberg 1985, 1986) in nature exhibit amphistomatic sun leaves.

Mott et al. (1982) concluded that placement of equal stomata on both epidermises, instead of limiting them to one surface, should maximize CO_2 uptake per unit area by doubling boundary layer conductance (3.1.5.1), and an amphistomatic leaf tends to have twice the number of stomata per leaf area, hence additional g^{st}. In the few critically analyzed nondesert amphistomatic species, the two conductances behaved in parallel when measured separately, as predicted (Mott and O'Leary 1984). For unstressed conditions, there was no significant difference in behavior of adaxial versus abaxial stomata over a wide range of external CO_2 concentrations; following water stress, adaxial conductance was more greatly reduced, but this condition was reversible.

Amphistomaty characterizes thin to thick nonsucculent desert leaves, but also C_4 leaves and sun leaves of nearly all herbaceous crops. This is consistent with the hypothesis that amphistomaty is primarily an adaptation to increase photosynthetic rate, and inconsistent with an earlier hypothesis (Parkhurst 1978), that suggested that amphistomaty is an adaptive design to shorten the CO_2 diffusion pathway, especially in thick leaves. Given that the rate for gas diffusion through a leaf theoretically decreases linearly with distance (Nobel 1991a), depending on other conductance terms in the diffusion pathway, some slowing of photosynthesis can be predicted as leaf thickness increases and so also resistance to a point where that is the limiting parameter in photosynthesis. Thus, amphistomaty is expected for very thick nonsucculent and succulent photosynthetic organs. Amphistomaty may reduce WUE (Foster and Smith 1986).

Guard cell length is relatively short, ranging from 46–17 µm in North American species of lowland desert scrub, and stomatal pores tend to be slightly longer on the adaxial surface. Short guard cells should be expected inasmuch as groundmass cells also have comparatively small volumes, but correlations of the two parameters were not statistically significant. Stomatal pore length was 45–55% that of guard cell length.

Since Maximov (1929, 1931), it has been known that leaves of well-watered desert perennials can exhibit high transpiration rates, comparable with those of some crop plants. Combining many small-aperture stomata from both surfaces can yield maximal conductances above 10 mm s^{-1} for deciduous leaves of many well-watered desert species and slightly lower values for desert evergreens, such as *Larrea tridentata* (Cunningham and Strain 1968; Strain 1970; Mooney et al. 1976a). These observations emphasize that the anatomical design of a desert leaf is primarily first to permit the highest possible CO_2 conductance, whenever soil water is available, and this is why the surfaces of these desert leaves do not have the expected features for conserving water, which would have the consequence of lowering stomatal conductance.

Because a wide pore on average is predicted to have a lower WUE than a very narrow one, given that reducing pore diameter increases stomatal resistance to water vapor more than for CO_2, many small pores theoretically have greater WUE than fewer wide pores at the same conductance. Hence, the apparent pattern of having small epidermal cell volumes may simply be a correlation with small stomatal pores and its water economy implications, instead of an adaptation to maintain turgor or water content in groundmass epidermal cells, which have extremely low water vapor conductances because of their cutinized walls.

Deeply sunken stomata may occur but are extremely rare among nonsucculent leaves of woody desert plants, and if sunken stomatal pores rarely are more than 10 μm (Fig. 3.7A–E), or approximately half the thickness of groundmass epidermis. Guard cells of the evergreen *Zygophyllum prismatocarpum* are sunken to 20 μm, less so in fleshy leaves of *Z. dumosum* (Evenari 1938b), and those of *Capparis aegyptia* may be sunken 50 μm (Fig. 3.2D; Volkens 1887). In North America the notable exceptions are the vestigial leaves of *Glossopetalon nevadensis* and *Castela polyandra* (having multiple epidermis), with stomata recessed more than 25 μm; these same species are aphyllous shrubs, having assimilatory stems with deeply sunken stomata, suggesting that those on the leaves are incidental.

As discussed earlier, calculations show that shallowly sunken stomata, as are found in *Moltkiopsis ciliata* and *Stanleya pinnata* (Figs. 3.7E), should have no physiologically significant effect on the conductance pathway and therefore would not reduce transpiration (Gibson 1983). A sunken stoma would theoretically cause a lowering of g_{wv} only if the pore was extremely deep, and should have virtually no effect during drought. Considering these results, a physiologist would conclude that nonsucculent desert leaves should not have sunken stomata if the adaptive strategy is to maximize CO_2 uptake, as suggested by amphistomaty.

Among desert leaves, stomatal crypts, abaxial stomatal furrows, and revolute margins, which characterize numerous species in mediterranean-type ecosystems, occur only on several of species with hypostomatic leaves, e.g., *Condalia* spp. and *Thymelaea hirsuta*.

3.2.1.3 Mesophyll

3.2.1.3.1 Thickness

In general, nonsucculent xerophytes are described as having thicker leaves than mesophytes, because they have more highly developed palisade parenchyma in

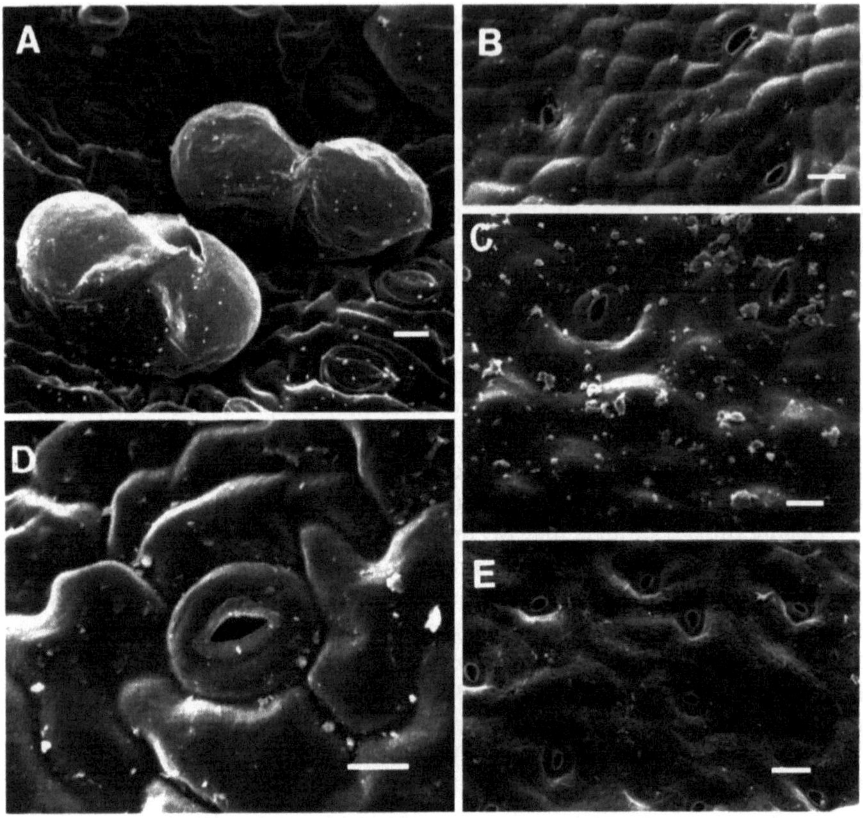

Fig. 3.7A–E. Stomata on adaxial leaf surface. **A** *Eriocephalus africanus*, where stomata are slightly recessed. **B** *Gomphocarpus sinaicus*, with shallowly sunken stomata. **C** *Rhazya stricta*. **D** *Rosenia humilus*. **E** *Stanleya pinnata*, with shallowly sunken stomata. *Bar* 10 µm

response to higher levels of PPFD (Nobel 1991a). Along similar lines, sun leaves have thicker mesophyll than shade leaves (Shields 1950), but the nature of PPFD exposure strongly affects the number of layers and cell shape (Chabot and Chabot 1977; Chabot et al. 1979; Jurik et al. 1979). On a single desert plant, leaves formed in cool months with abundant soil moisture may be half the thickness of leaves formed as the plant enters summer drought. To date, nonsucculent upper canopy desert leaves have been described with mesophyll ranging from 90–1300 µm, and from that sample, mean mesophyll thickness was 84% (range 66–94%) of intercostal laminal thickness (excluding trichomes).

3.2.1.3.2 Arrangement of Palisade Parenchyma

Typical desert leaves possess isolateral (unifacial to centric) mesophyll, having at least some palisade parenchyma on the abaxial side or, in many, palisade parenchyma throughout the transection (Fig. 3.8A–E) and frequently continu-

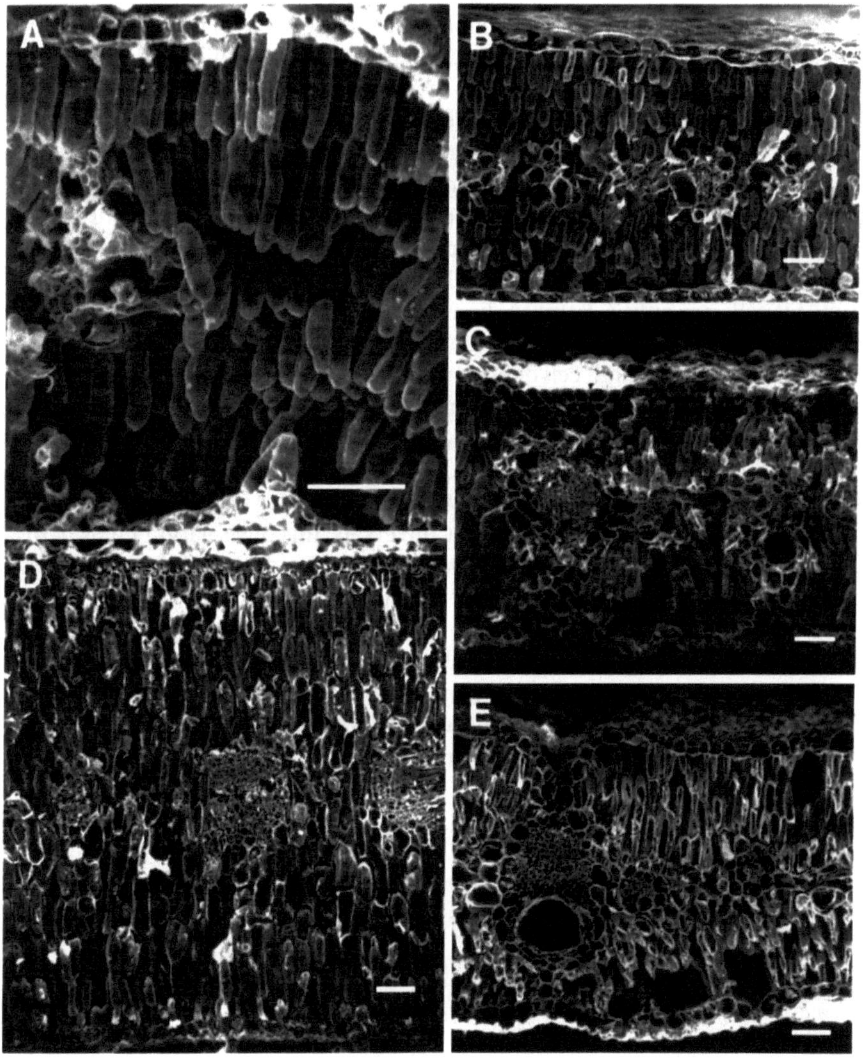

Fig. 3.8A–E. Leaves with isolateral mesophyll. A *Ambrosia deltoidea*. B *Crossosoma bigelovii*. C *Flourensia cernua*. D *Simmondsia chinensis*. E *Ericameria linearifolia*. *Bar* 50 μm

ing around leaf margins. On average, adaxial palisade is about 20% thicker than the abaxial zone, and abaxial cells tend to be shorter. In New World deserts, prominent exceptions, having strongly bifacial mesophyll, are hypostomatic leaves of *Berberis trifoliolata, Rhus microphylla*, and *Atamisquea emarginata*, none of which occurs in typical lowland habitats. Of species described from the Saharo-Sindian zone, *Chrozophora brocchiana* has bifacial leaf design (Stocker 1972) and *Ziziphus lotus* is diagrammed as bifacial (Stocker 1974a), but instead may be considered intermediate.

Descriptions of desert leaves emphasize that palisade parenchyma is dense and consequently forms small intercellular air spaces (Volkens 1887; Maury 1887; Sabnis 1919–1921; Shields 1950, 1951a). Each groundmass epidermal cell is associated with at least one palisade mesophyll initial, and as the epidermal cell enlarges, cell divisions are encouraged in the more cylindrical palisade cells. At maturity, a single epidermal cell may bear up to 12 narrow palisade cells (Fig. 3.9), accounting for the high cell density and small intercellular air spaces.

Viewed with SEM, paradermal sections through adaxial and abaxial mesophyll show high cell density (Fig. 3.10A–D), in some species exceeding 11 000 mm^{-2} if cells are only 7–8 μm in diameter, and fewer in those with wider palisade cells. Cell density in adaxial palisade always exceeds abaxial density by at least 20%. Extremely high cell density with narrow cells creates the image, using thick sections with the light microscope, of very tight packing, but SEM photomicrographs reveal that typically less than 10% of lateral walls remains connected (Fig. 3.10C–D). The fully exposed lateral walls, next to which are arranged many chloroplasts (Figs. 3.10B), should yield the highest possible rates of CO_2 uptake by chlorenchyma cells. Conversely, this observation helps to dispel the belief that dense mesophyll, with much crowding and thereby greater touching of cells, should be a desert-leaf adaptation to retard evaporation from internal surfaces (Sabnis 1919–1921). In contrast, recent studies using careful techniques demonstrated that internal evaporation is largely restricted to mesophyll cell adjacent to guard cells (Appleby and Davies 1983).

In each desert, there are a few exceptions, where palisade cells remain connected more than 50%, producing cylindrical and narrow intercellular air space.

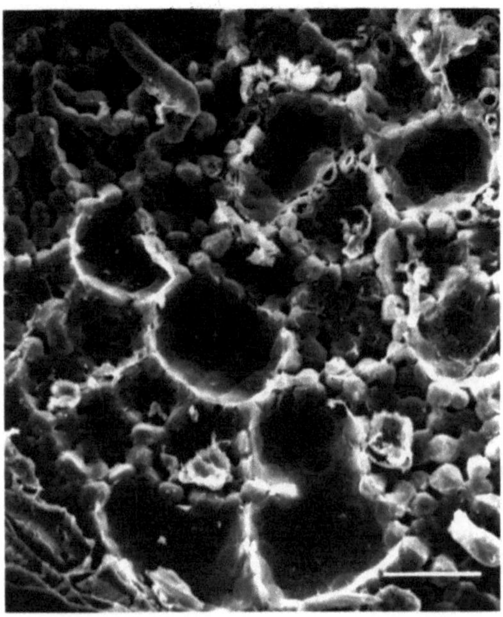

Fig. 3.9. *Sageretia wrightii*; paradermal view of palisade mesophyll cells attached to adaxial epidermal cells. *Bar* 25 μm

This condition is most common for thick leaves, such as in *Simmondsia chinensis* (Fig. 3.8C), *Mortonia sempervirens, Maerua crassifolia* (Stocker 1974a), and species of *Capparis*, and it can be predicted that these should have tougher leaves and relatively low photosynthetic rates, because liquid-phase CO_2 has less surface through which to diffuse. A leaf having very high attachment of palisade cells theoretically could yield a longer and perhaps more tortuous pathway for CO_2 to diffuse from stomatal pore to chlorenchyma wall (Parkhurst 1978). Probably, leaves with highly touching mesophyll cells also have thicker walls, which impair gas diffusion.

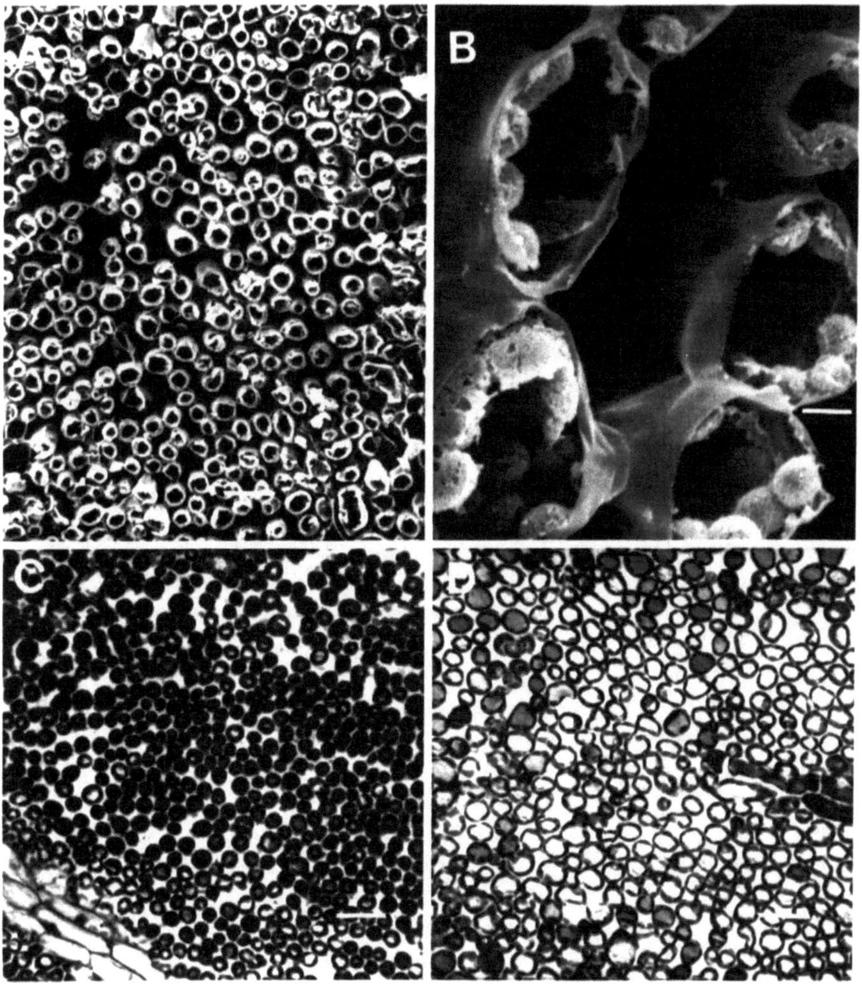

Fig. 3.10A–D. Paradermal views of adaxial palisade mesophyll. **A** *Chilopsis linearis.* **B** *Gomphocarpus sinaicus.* **C** *Ambrosia cordifolia*, light microscope section to show low percentage of touching for palisade mesophyll cells. **D** *Flourensia cernua. Bar* **A,C,D** 25 μm; **B** 5 μm

Hypodermis, having either collenchymatous or sclerified cell walls, is virtually unknown among leaves of nonsucculent desert plants. Notable exceptions are several evergreens, e.g., *Maytenus polyacantha* (Fig. 3.6B), which has a thick-walled hypodermis with included druses, and *Simmondsia chinensis*, which has a weakly defined hypodermal layer (Fig. 3.7C). Hypodermis appears in *Coleogyne ramosissima* from relatively cool habitats.

3.2.1.3.3 Substomatal Chambers

During leaf differentiation, no mesophyll cell remains attached to the guard cell mother cell, so that when the pair of guard cell differentiates a space, the substomatal chamber (Fig. 3.11A–D), is formed internal to the stomatal complex, one palisade cell deep. Much attention has been given to this region, because

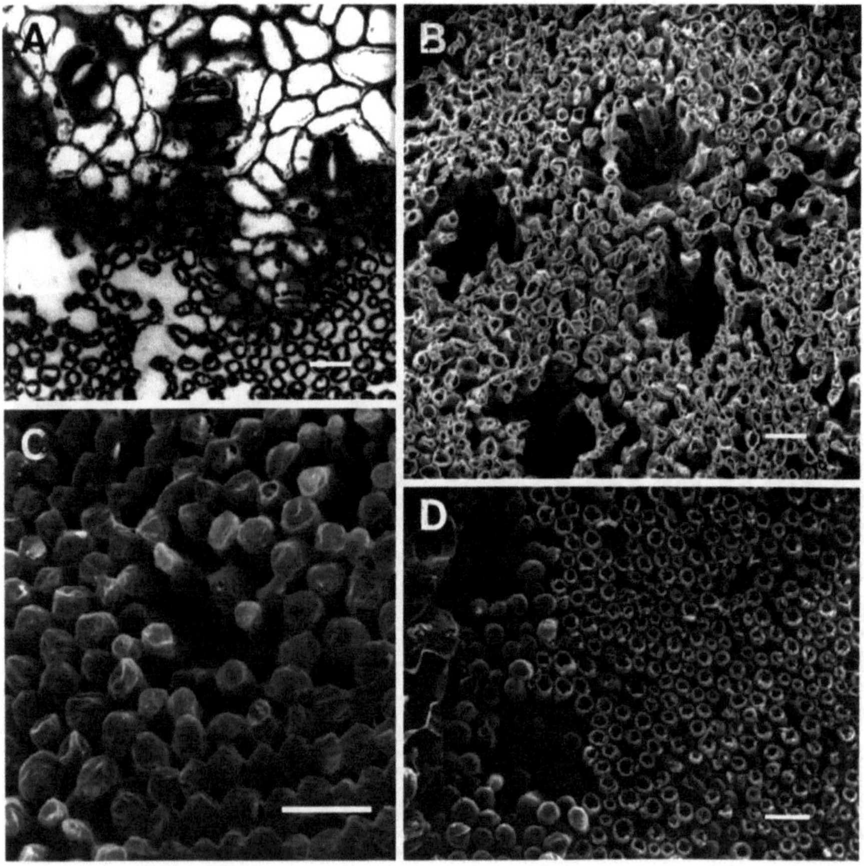

Fig. 3.11A–D. Paradermal views showing substomatal chambers, which create a prominent intercellular air space interior to each pair of guard cells. **A** *Chilopsis linearis* (*arrow*). **B** *Mortonia sempervirens*. **C** *Becium burchellianum*. **D** *Crossosoma bigelovii*. *Bar* 25 μm

desert leaves tend to have smaller substomatal chambers than mesophytes. Small chambers result from having short guard cells. It does not appear likely that the size of the substomatal chamber would affect either transpiration or photosynthetic rate other than by modifying the linear distance from the pore to the mesophyll cell, presumably a physiologically insignificant addition to the pathway.

Substomatal chamber is, in fact, a confusing misnomer in these and other photosynthetic organs. Chamber connotes the presence of a system of walls that mostly encloses the space. SEM photomicrographs of desert leaves reveal that the space is, instead, defined by columnar cells with wide intercellular air spaces between them (Fig. 3.11B,C), and there probably is no impairment for diffusion of CO_2 to palisade cells.

3.2.1.3.4 Wall Thickness

Liquid-phase diffusion of CO_2 is relatively rapid through thin chlorenchyma cell walls (3.1.4.5) but would be slowed by the presence of thicker walls. All mesophyll parenchyma have relatively thick primary cell walls in evergreen leaves of *Mortonia sempervirens*. Spongy mesophyll has thickened walls in several Asteraceae with ericoid leaves, e.g., *Chrysothamnus paniculatus* and *Ericameria linearifolia*. In *Ambrosia dumosa*, a drought-deciduous chamaephyte of the California deserts, the relatively thick leaf (mean 470 µm) appears to be partitioned into separate halves by a prominent spongy mesophyll having lignified cell walls (Fig. 3.12). Spongy mesophyll in desert leaves has been classified as water reservoir (Shields 1951a) or water-storing tissue, but it is uncertain whether these cells have fewer chloroplasts because they are specialized for water storage or because PPFD was insufficient, or diffusion pathway length was too great, to utilize the central cells for photosynthesis.

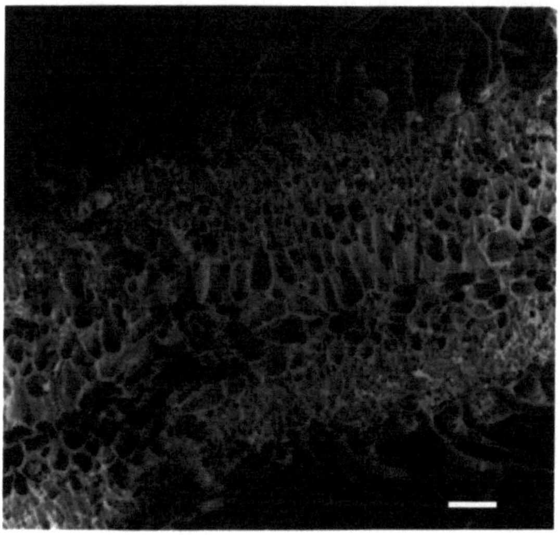

Fig. 3.12. Leaf transection of *Ambrosia dumosa* showing reflective trichomes on both surfaces; adaxial and abaxial palisade are separated by central mesophyll with lignified cell walls. *Bar* 100 µm

3.2.1.3.5 A^{mes}/A

Very few accurate estimates of A^{mes}/A exist for leaves from warm deserts, excepting those published on *Hyptis emoryi* grown under different PPFD regimes, which were 14–39 from one-sixth sun to full sun (Nobel 1976a; Smith and Nobel 1977b), and *H. emoryi*, *Encelia farinosa*, and the herbaceous perennial *Mirabilis tenuiloba* sampled from field plants throughout the year (Smith and Nobel 1977a). Typical values of A^{mes}/A for wild-collected, full-sun desert leaves are 25–40, but many species need to be examined to determine the means and patterns for evergreen and deciduous leaves and from different deserts. Because adaxial palisade is thicker and denser, adaxial A^{mes}/A always exceeds the abaxial one. Nonsucculent desert leaves can have relatively high A^{mes}/A values, and therefore can experience photosynthetic rates similar to those of row crops (Nobel 1991a).

Many of the thinnest desert leaves (< 200 µm, Table 3.2), hence most likely those with relatively low values of A^{mes}/A, are leaves that have a dense, reflective layer of trichomes on each surface (Fig. 3.13A,B).

3.2.1.3.6 Idioblasts and Special Inclusions

Desert leaves are noteworthy for the general absence of unusual idioblasts, other than those of phylogenetic significance, such as myrosin cells in Brassicaceae, cystoliths in Ulmaceae, and tannin cells in Rosaceae, Rhamnaceae, and Simmondsiaceae (Metcalfe and Chalk 1989). Crystaliferous cells are most abundant in evergreens, excepting *Larrea*, and noticeably scarce in most nonsucculent deciduous leaves except adjacent vascular bundles (Sabnis 1919–1921), so it is not likely that they are more abundant than crystals in other vegetation

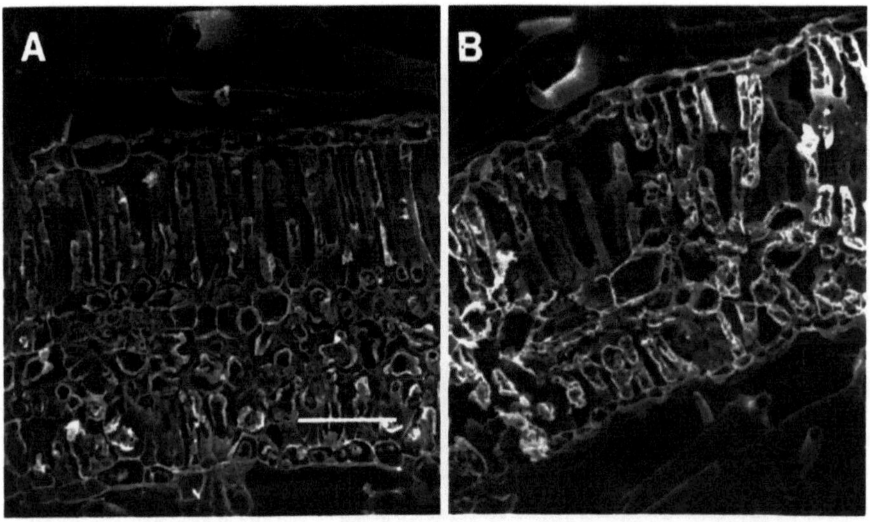

Fig. 3.13A,B. Leaves with trichome-shaded epidermis, and palisade parenchyma is weakly expressed in abaxial mesophyll. **A** *Stachys aegyptiaca*. **B** *Chrozophora obliqua*. *Bar* 25 µm

types. Crystal deposition increases with leaf age, therefore evergreen leaves would tend to accumulate mesophyll crystals.

Another indication of crystal increase with age is the occasional occurrence of crystal aggregates with substomatal chambers, plugging stomata, to date observed in such diverse taxa as *Ceraria namaquensis* (Michell 1912), *Crossosoma bigelovii* (Richardson 1970), *Zygophyllum prismatocarpum*, and *Capparis aegyptia*. These crystal aggregates also develop in adjacent epidermal and palisade cells. Ontogeny of extracellular calcium deposits has not been described for any desert species, but seems to occur for plants grown on soils rich in calcium minerals, especially limestone and gypsum. In these same leaves, crystal aggregates also can precipitate within and thereby plug vessels of leaf veins.

3.2.1.3.7 Vascular Sclerenchyma

Traditional summaries on xeromorphism emphasize the high percentage of species having sclerenchyma associated with vascular bundles (Fahn and Cutler 1992); but fully differentiated leaves of most woody desert species rarely have either fibers or sclereid idioblasts.

In typical warm deserts, evergreen leaves tend to have some fibers associated with the primary and largest secondary veins, as in the North American examples *Mortonia sempervirens* (midvein only), *Simmondsia chinensis*, *Berberis* spp., *Viscainoa geniculata*, and *Larrea tridentata*. In Zygophyllaceae, in addition to *Larrea* and *Viscainoa*, the drought-deciduous shrub *Porlieria angustifolia* has massive clusters of fibers associated with each vascular bundle, and even chamaephytes of *Fagonia* have some bundle fibers. Among Asteraceae, midveins in less than half the warm desert species have some primary phloem fibers, but the most fibrous leaf observed was the winter-deciduous *Artemisia tridentata* of cool desert habitats (Shultz 1986). Very thick-walled fibers occur in some ephemeral leaves, e.g., next to protoxylem in *Castela*. Microphylls also show low frequency of midvein collenchyma and bundle sheath extensions.

Major exceptions to an overall absence of sclerenchyma in woody plants come from Australia, where sclerophylly is common across the continent, even in rainforest canopies associated with low soil fertility, especially phosphorus starvation (Beadle 1966). In central Australia species of *Hakea* that have sclereids interrupting the palisade mesophyll of terete adult leaves with sunken stomata have flat juvenile, amphistomatic, isolateral leaves without sclerenchyma and having superficial stomata (Vischer 1915).

3.2.2 Phyllodineous Acacias

In *Acacia* of Australia, narrow to terete phyllodes occur on the species of the arid habitats, belonging to sections *Juliflorae*, *Plurinerves*, and *Phyllodineae*.

Acacia aneura, a very widespread species, has epidermis consisting of club-shaped cells with heavily cutinized, thickened walls. Stomata are arranged in longitudinal rows and sunken up to 25 µm. Palisade chlorenchyma of this species has been described as interrupted, i.e., having sectors of chlorenchyma alternating with sectors of primary phloem fibers and crystaliferous cells, but in other aridland species fibers are absent (Boughton 1986). Palisade cells are narrow, and a ring of vascular bundles encloses central mesophyll, consisting of water-storage parenchyma with thin walls.

3.2.3 Woody Dicotyledons with C_4 Photosynthesis

Relatively few species of leafy, nonsucculent woody plants of warm deserts exhibit C_4 photosynthesis. The majority belong to the family Chenopodiaceae, including obligate and facultative halophytes.

Although certain studies have provided careful descriptions of only desert taxa (e.g., Volkens 1887; Evenari and Richter 1938; Zemke 1939; Black 1954; Osmond 1974), leaf anatomy for many others can be found embedded in larger treatments of Chenopodiaceae (Carolin et al. 1975, 1978, 1982), recognized as atriplicoid, kochioid, and salsoloid Kranz types (Fig. 3.14A–C). Desert species are representatives of all three types and differ in no obvious ways. All C_4 species

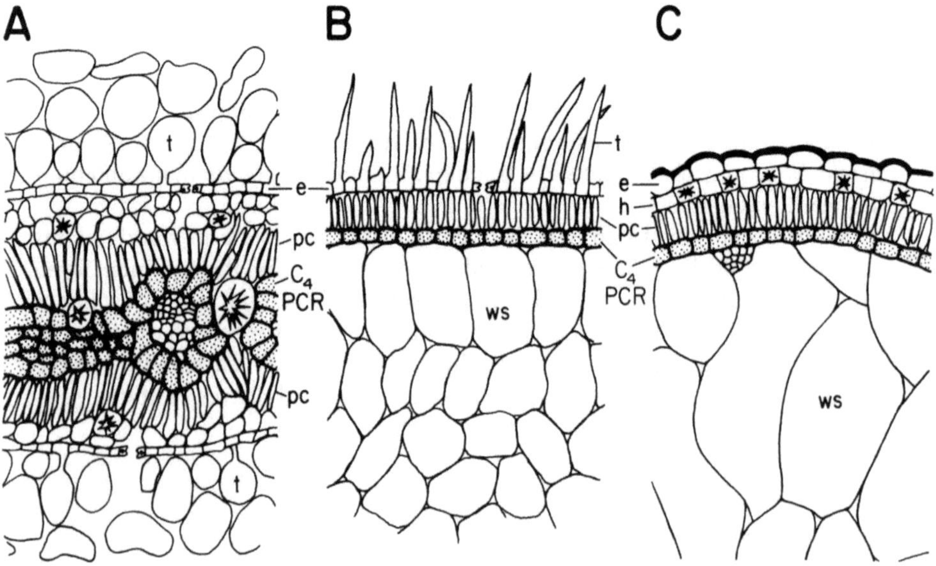

Fig. 3.14A–C. Transections of C_4 leaves for various Chenopodiaceaae. **A** *Atriplex barclayana.* **B** *Maireana sedifolia.* **C** *Salsola longifolia.* (**C** after Volkens 1887). *Stippled layer* is the C_4 PCR tissue. *e*, epidermis; *h*, hypodermis; *pc*, palisade chlorenchyma; *t*, trichomes. *Bar* 50 µm

that have been described are amphistomatic, with similar stomatal densities on both surfaces.

Leaves of the desert *Atriplex* have relatively thin-walled epidermis and thin-walled, highly vacuolate, adaxial and abaxial hypodermis touching radially arranged palisade mesophyll associated with each vascular bundle (Evenari 1938b). Collenchyma tends to occur along the midvein, extending from the stem cortex through the short petiole.

3.3 Phreatophytes

Leaves of desert phreatophytes may be microphyllous to extremely wide, but characteristically they are *water-spenders* (sensu Levitt 1980), having high stomatal conductance, thereby utilizing transpiration to cool leaves below lethal temperatures and simultaneously experiencing moderate to high rates of carbon assimilation for high biomass accumulation (Ludwig 1987). A microphyllous example is *Acacia raddiana*, with daily summer transpirational losses of 2.9 kg H_2O m^{-2}, and a macrophyllous example is *Calotropis procera*, with losses of 2.3 kg H_2O m^{-2} but with less than one-third of the daytime CO_2 uptake (Stocker 1971).

3.3.1 Desert Palms

Within the family Arecaceae, which mostly has hypostomatic leaves (Tomlinson 1961), species of desert palms, inhabiting oasis woodland and springs (Chap. 1), exclusively have amphistomatic leaves, and stomata occur at relatively high densities on any leaf that differentiates in full sun. Four species surveyed had equal numbers on adaxial and abaxial surfaces of the same leaf, whereas five had significantly higher densities of abaxial stomata. Summed from the two surfaces, several species have >550 stomata mm^{-2}; these are remarkable values given that epidermal strips opposite vascular bundles and their sclerenchymatous girders are astomatic, thereby eliminating half of each surface from functioning in gas exchange.

Leaves are glabrous, and primary cuticle tends to be thick on palm leaves, frequently covered by abundant epicuticular wax, which almost appears to plug stomata in several species. Among the species examined were species with straight anticlinal epidermal cell walls (*Trithrinax campestris*) versus undulating ones (*Livistona mariae*), and conspicuously sunken, large stomata (*T. campestris*) versus shallowly sunken, small stomata (*Sabal uresana*). These palms do not have palisade chlorenchyma. Anatomy of a mature leaf of *Hyphaene thebaica* has been schematically illustrated (Fahn and Cutler 1992).

Abundant sclerenchyma associated with vascular bundles certainly functions to strengthen the large leaf.

3.3.2 Tropical Dicotyledonous Trees

Many phreatophytes of warm deserts are legumes, particularly deciduous species of *Acacia* (Fig. 3.15A) and *Prosopis* with microphyllous, amphistomatic, isolateral leaflets (Table 3.2). Relatively thick primary cuticle is common in *Prosopis* (Bleckman and Hull 1975; Mooney et al. 1977b), and species often have well-developed clusters of primary phloem fibers (Fig. 3.15B), which may enable leaflets to resist damage from large diurnal differences in water potential (Szarek and Woodhouse 1976, 1977).

Salvadora persica is a small tree or shrub with evergreen leaves that show no water economy while living under extremely arid conditions (Stocker 1970, 1971; Sen et al. 1972; Evenari and Gutterman 1973). These leaves are 300–350 μm thick, amphistomatic, with similar stomatal densities on both surfaces, and

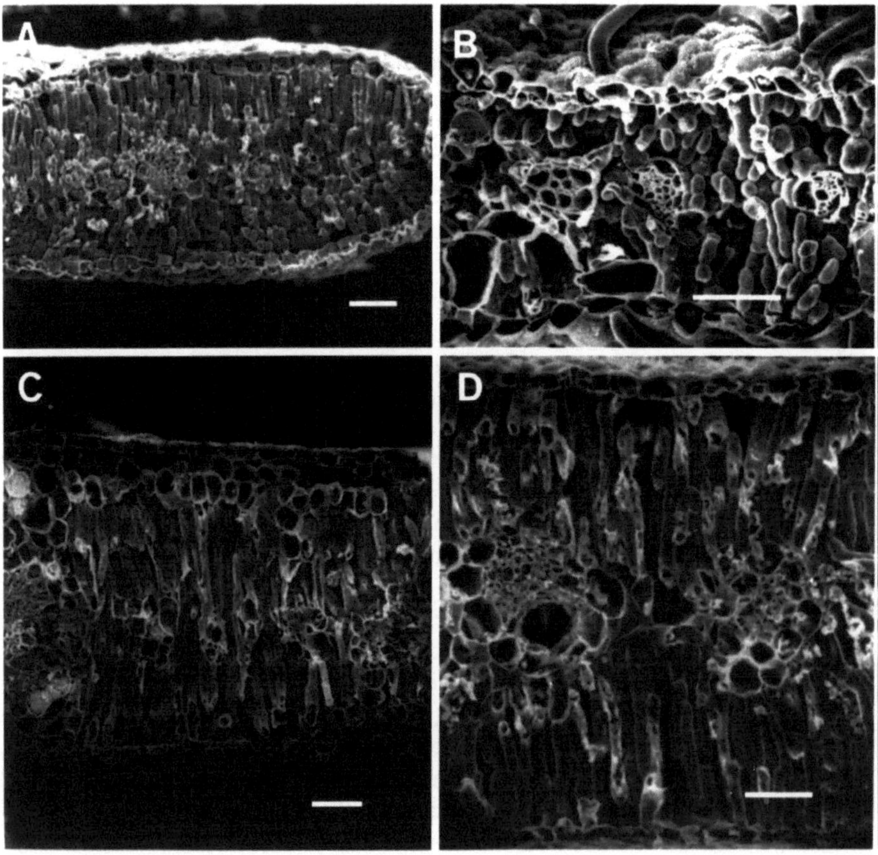

Fig. 3.15A–D. Leaf transections of desert riparian perennials. **A** *Acacia raddiana*. **B** *Olneya tesota*. **C** *Populus fremontii*. **D** *Baccharis emoryi*. *Bar* 50 μm

isolateral, and the biseriate epidermis has thick cell walls (Pillai and Pillai 1977). Leaf temperature remains close to ambient temperature, and leaf water content can drop on a hot day to below 80% by experiencing transpiration rates of 5.4 $\mu g\ mm^{-2}\ s^{-1}$ (Sen et al. 1972).

From North America, *Chilopsis linearis*, a deep-rooted small tree with pendent, somewhat glutinous, amphistomatic, isolateral leaves (Scott 1935a; Henrickson 1985), in the field never experiences leaf water potentials lower than -1.7 MPa, and appears to have a favorable WUE while maintaining relatively good photosynthetic rates during drought (Odening et al. 1974).

Calotropis procera has amphistomatic leaves with high stomatal densities but relatively small pore lengths (Stocker 1971).

3.3.3 *Welwitschia mirabilis*

Welwitschia mirabilis, a relict gymnosperm from the fog zone of the Namib, has a single pair of extremely wide, fibrous and curled, persistent adult leaves up to 1.3 mm thick. Each leaf grows at the base via an intercalary meristem and has tattered tips with extensive basipetal splitting. Injured portions of leaves produce wound periderm (Chamberlain 1935; Salema 1967).

Although leaves of *Welwitschia* contain relatively high levels of organic acids, typical of CAM plants, daytime CO_2 uptake has been found in field plants, indicating that this really is a C_3 species (Schulze et al. 1976b; Schulze and Schulze 1976; von Willert et al. 1982).

Epidermal cells have thick and cutinized outer periclinal walls, also possessing crystal sand, and covered with a thin primary cuticle to 3 µm. Anticlinal and inner periclinal walls are also thickened, indicating an adaptation for minimizing cuticular transpiration. Stomata, which are amphistomatic and sunken to 30 µm, are dense (approximately 150 mm^{-2}), but absent opposite sclerenchymatous girders, and pores are oriented only longitudinally, as in desert palms. Leaves are highly reflective, which aids in limiting heating of these broad leaves, and, remarkably, shade cast by these leaves also figures significantly in cooling full-sun leaves (Schulze et al. 1980).

Leaves are isolateral, having three to four palisade layers (200–250 µm) on each side of central mesophyll, composed of isodiametric parenchyma (Rodin 1958a, 1958b; Napp-Zinn 1966; Butler et al. 1973). Palisade chlorenchyma occurs in longitudinal strips because the tissue is subdivided by parallel clusters of unlignified hypodermal fibers. Mesophyll of *Welwitschia* also contains large, branched sclereid idioblasts that possess crystal sand of calcium oxalate between the primary and secondary cell walls. No water storage occurs in the leaf, which typically has only 45–65% water content (von Willert et al. 1982).

Leaf vasculature consists of large longitudinal veins and smaller oblique anastomosing bundles. Principal veins have primary fibers associated with phloem and xylem, and old vascular bundles are also surrounded by sclerified parenchyma of central mesophyll.

3.3.4 Temperate Trees and Shrubs

Important phreatophytic trees are broad-leaved species of *Populus* and *Salix*, which are obligate phreatophytes that generally use only groundwater (Smith et al. 1996). These species are mesophytes in most senses, but it is worth noting that full-sun leaves are amphistomatic and have high A^{mes}/A ratios with strong development of palisade mesophyll (Fig. 3.15C), required for high transpiration losses for cooling the broad leaves and likewise designed to provide high stomatal conductance for CO_2 uptake.

Among shrubs growing at springs or wadis are temperate groups with *Salix*-like leaves, e.g., North American species of *Baccharis* and *Pluchea sericea*. The lanceolate leaf design and flexible shoots are assumed to be adaptations for minimizing mechanical damage during floods. These leaves are amphistomatic and have isolateral mesophyll (Fig. 3.15D).

3.3.5 Saltcedars and Casuarina

Saltcedars (*Tamarix*), originally from Asian deserts, and casuarina or desert she-oak (*Casuarina descaineana*), endemic to near-arid central Australia, are trees with highly branched and very thin terminal green shoots having minute leaves. Several species have been introduced around the world in arid and semiarid regions.

In *Tamarix*, such as the naturalized riparian trees *T. ramosissima* of North American and *T. aphylla* of the Negev, leaves form in a condensed phyllotactic helix, but as shoots age a leaf enlarges to several millimeters in length (Fig. 3.16A), and cauline leaves may be 8–9 mm in length (Wilkinson 1966). These minute, thin leaves, which remain appressed to stems, have transverse stomata mostly on the abaxial surface facing the stem and shallowly sunken. Microphylls have palisade mesophyll but progressively experience subepidermal wax deposition. Thin stems are yellowish green due to the presence of chloroplasts in isodiametric cells of a thin outer cortex, but there are so few stomata on stems that a stem is not an efficient photosynthetic organ. On young stems, epidermis has extremely thick and heavily cutinized outer periclinal cell walls and thick anticlinal and inner periclinal walls (Fig. 3.16B). Seedling leaves of *Tamarix* tend to lack a sharply defined palisade layer (Wilkinson 1966).

Casuarina descaineana, like all Casuarinaceae, has whorled microphylls (Dilcher et al. 1990); *Oberblatts* are free at the node, whereas *Unterblatts* are decurrent leaf bases that become distinctive green vertical ridges of each inter-

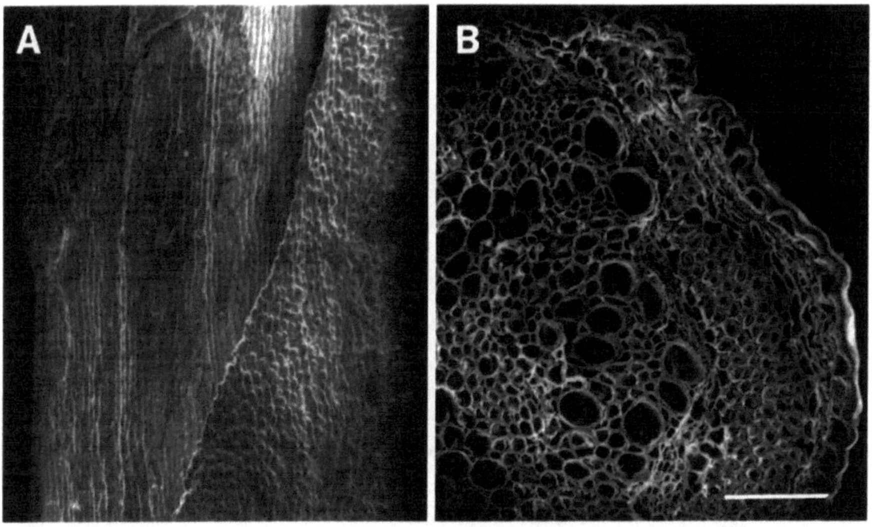

Fig. 3.16A,B. *Tamarix ramosissima.* **A** Photosynthetic microphyll (*right*) and nonphotosynthetic stem. **B** Stem transection, having heavily cutinized epidermis but no stomata or palisade chlorenchyma. *Bar* 50 μm

node (Fig. 3.17A). The central ridge line, corresponding to the leaf midvein, is reinforced with a T-shaped girder of fibers, eventually connecting epidermis to phloem of the leaf-trace bundle. On either side of the girder occurs several-layered palisade parenchyma (mesophyll) to 125 μm in thickness (Fig. 3.17B; Metcalfe and Chalk 1950). Transverse stomata are present on the sides of furrows, adjacent palisade cells, not on the ridges, where epidermis has thick walls with micropapillae.

3.4 Herbaceous Species

3.4.1 Perennial and Annual Grasses

Considering grass floras of warm deserts, nearly all of the ecologically important bunchgrasses are C_4 species, derived from moister subtropical and tropical floras. In the low-elevation Monte, C_4 species comprise greater than 90% of grasses present (Cavagnaro 1988), and the warm desert grasses of Australia comprise up to 100% C_4 species in semiarid to arid regions where a summer rainfall regime dominates (Hattersley 1983). As defined in this monograph, the desert zone of Australia has approximately 160 species of C_4 grasses, especially the dominant sclerophyllous bunchgrasses of *Triodia* and *Plectrachne* and *Zygochloa paradoxa* on unstable dune crests (Prendergast 1989). In Africa,

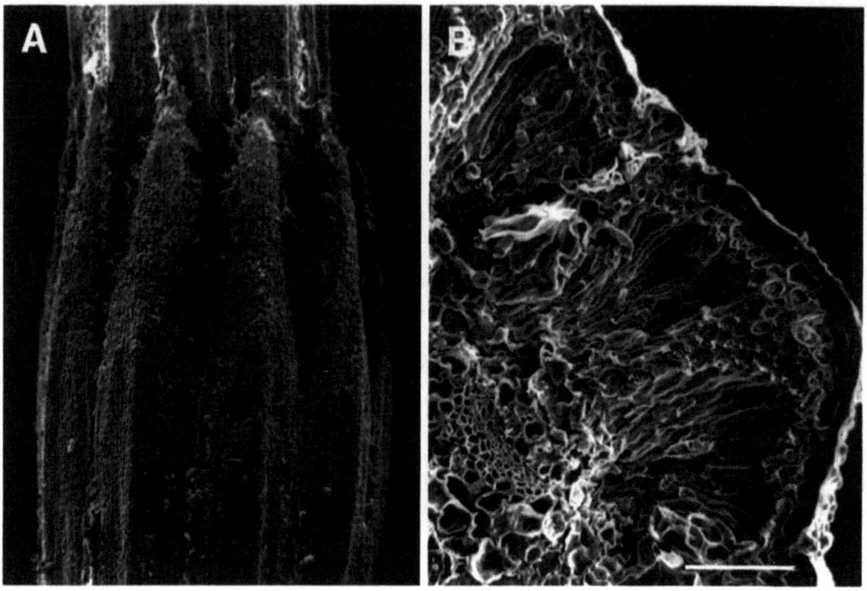

Fig. 3.17A,B. *Casuarina* aff. *descaineana*. **A** Free tips of microphylls, which are adnate to the stem. **B** Transection showing that leaves form the photosynthetic portion of the axis. In **B** stomata are positioned in grooves (leaf margins), palisade parenchyma comprise leaf mesophyll, and girdles of primary phloem fibers form opposite the leaf vascular bundle. *Bar* 50 μm

successful colonizers of sand dunes are rhizomatous perennial species of *Stipagrostis* (tribe Aristideae) and *Panicum turgidum* (tribe Paniceae), C_4 grasses.

Desert C_4 grasses include examples of acid decarboxylation types NAD-malic enzyme (NAD-ME), NADP-malic enzyme (NADP-ME), and phosphoenolpyruvate carboxykinase (PCK). Leaves having a "single sheath" (XyMS−) of C_4 photosynthetic carbon reduction (PCR) tissue are NADP-ME type (Fig. 3.4), with centrifugal chloroplasts and uneven sheath outline (Hattersley and Watson 1976; Dengler et al. 1985; Prendergast and Hattersley 1987); this characterizes common desert species in genera of subfamily Arundinoideae, including *Eriachne* and *Stipagrostis*, but the closely related *Aristida* has a unique double PCR sheath (Johnson and Brown 1973).

Leaves having a double sheath (XyMS+), with an outer PCR sheath and inner mestome sheath, are of the PCK type, where PCR bundle sheath outline is uneven and PCR chloroplasts are centrifugally arranged in cells with suberized lamella, as in *Chloris*, *Pleuraphis*, and many species in *Sporobolus* and *Bouteloua* of subfamily Chloridoideae. In Chloridoideae with even bundle sheath outline and centripetal PCR chloroplast arrangement, the NAD-ME type occurs, e.g., *Eleusine*, *Distichlis*, and *Astrebla*. However, for arid zone chloridoids there are examples of NAD-ME types in which chloroplast position is variable, e.g., *Eragrostis* and *Enneapogon*, and *Plectrachne schinzii* and closely related *Triodia* (Burbridge 1946) have centrifugal chloroplasts. Probably *Erioneuron* also belongs to the NAD-ME type.

Within subfamily Panicoideae, XyMS – and XyMS+ as well as all three enzyme types appear, even within *Panicum*. Tribe Andropogoneae, which includes *Cymbopogon*, consistently has XyMS- and NADP-ME features, as do *Pennisetum* and *Cenchrus* (Paniceae), but the widespread *Panicum turgidum* has the NAD-ME type.

Among the C_3 desert taxa, which typically grow when watered during cool or mild seasons, are *Schismus, Centropodia, Asthenantherum*, and *Dregeochloa* (Arundineae, Arundinoideae), as well as *Stipa* and *Achnatherum* (Stipeae) and numerous annual and perennial genera of Pooideae that range into desert lowlands. These may have a single or double bundle sheath.

Typically, C_4 desert grasses have amphistomatic leaves, except on epidermal strips touching sterome fibers, whereas spring-active C_3 species show a greater tendency for stomatal distributions only on one side of the leaf. Stomatal densities and distribution patterns of desert grasses have not been reviewed.

Leaves of desert Poaceae, like those from other habitats, often contain a high percent of cell wall, either as sclerenchyma or sclerified parenchyma (de Winter 1965). Thick, lignified walls characterize the epidermis but preferentially in the abaxial layer, which can be a major mechanical feature in leaf involution or convolution (Chap. 2). Abaxial and adaxial sterome fiber strands and girders are commonly observed in leaves of desert species, undoubtedly designs to oppose mechanical stresses that for none of these have been quantified; hence, to date, discussions of fibers have been limited to utilizing these features for systematic treatments, or, as in the cases of spinifex grasses in Australia, also related to nutrient-poor soils (Beadle 1966). Amount of sclerenchyma is not strictly related to leaf longevity, because short-lived dwarf annuals like *Stipagrostis subacaulis*, which emerges on Namib sand dunes after only a light rainstorm, may form large sterome girders, whereas perennials such as *S. dinteri* may have very small ones (de Winter 1965).

Leaf reduction is achieved in *Stipagrostis* by disarticulation of the blade, leaving leaf sheath, possibly culm, to perform photosynthesis (de Winter 1965). Culm photosynthesis versus that of the persistent sheath has not been structurally and biochemically analyzed in desert taxa, but stem photosynthesis is most likely not significant, given that cortical chlorenchyma is never highly expressed, stomatal densities are low or zero, and leaf sheaths tightly envelop the stem and thereby provide a poor microclimate for CO_2 uptake.

3.4.2 Herbaceous Dicotyledons

Annuals and herbaceous perennials constitute the ephemeral florulas that have been so widely discussed as "drought-avoiding" and therefore presumably lacking structural adaptations to cope with physiological stresses of the otherwise arid environment. More correctly stated, nonsucculent leaves of desert herbs have structural designs and biochemical adaptations to maximize CO_2 uptake during the short aboveground growing phase of the life cycle, even if doing so involves large water vapor losses. Cool-weather warm desert annuals, those that experience shoot growth during late fall to mid-spring, are with few exceptions C_3 species, whereas these that must grow rapidly following rains in hot summer months are predominantly C_4 species (Mulroy and Rundel 1977). In some families, phylogenetic evidence is clear that the dwarf annual is an adaptation from perennial ancestors to avoid drought (Ellis 1984), but comparative physiological studies are needed to determine how the photosynthetic apparatus has evolved.

Most generalizations about deciduous leaves of woody desert perennials can be extended to include herbaceous C_3 species of full-sun microhabitats, although surprising little literature exists on the subject (Volkens 1887; Sabnis 1919–1921). Leaves are amphistomatic, often with relatively high adaxial and abaxial densities of small stomata that are not sunken (Fig. 3.18A–D). Trichomes may be present or absent, but cuticle is thin and most likely not as effective in reducing g_{wv} as that on leaves of sympatric perennial dominants, given that ephemerals have a tendency to wilt rapidly after being picked. Mesophyll is isolateral, sometimes with high density of palisade cells, e.g., the thick basal rosette leaves of *Eriogonum brachypodum* (Fig. 3.19A). Many microphyllous winter annuals have relatively large intercellular air spaces and relatively short, almost cuboidal palisade cells (Fig. 3.18B; Volkens 1887; Maury 1887). No study has analyzed mesophyll structure of desert annuals as it would relate to the PPFD of the season during which the leaves were produced, although in *Camissonia claviformis*, an annual with relatively large spring leaves, A^{mes}/A was 41 and CO_2 uptake was an astounding 74 µmol $m^{-2} s^{-1}$ (Longstreth et al. 1980). Other desert species of *Camissonia* appear to have relatively high values of A^{mes}/A but wide spacing of palisade mesophyll cells (Fig. 3.19C,D).

Winter desert ephemerals do not constitute a uniform type, as might be concluded from the literature. Individuals growing in full sun tend to have significantly higher leaf specific weights than those growing in shady microhabitats (Wallace and Szarek 1981). *Oxytheca perfoliata*, a diminutive annual of California deserts, has minute, amphistomatic basal microphylls with low stomatal densities and very weak development of palisade but conspicuous, waxy, amphistomatic photosynthetic bracts on the inflorescence, with much higher stomatal densities and well-developed adaxial and abaxial palisade. *Chorizanthe rigidus* also has floral amphistomatic floral bracts with high A^{mes}/A.

Among Asteraceae, Kranz anatomy and carbon isotope ratios together indicate that only *Pectis* of subtribe Tageteae are summer-active C_4 desert species, as are several aridland species of *Flaveria* in subtribe Heleneae, both cases which have C_3 ancestors (Smith and Turner 1975). Some C_4 perennial dicotyledons, such as species of *Chamaesyce* and *Tidestromia*, begin active growth before mid-spring but are exposed to high temperature near ground level. A desert like the Thar, with predominantly summer rainfall, correspondingly has seve-

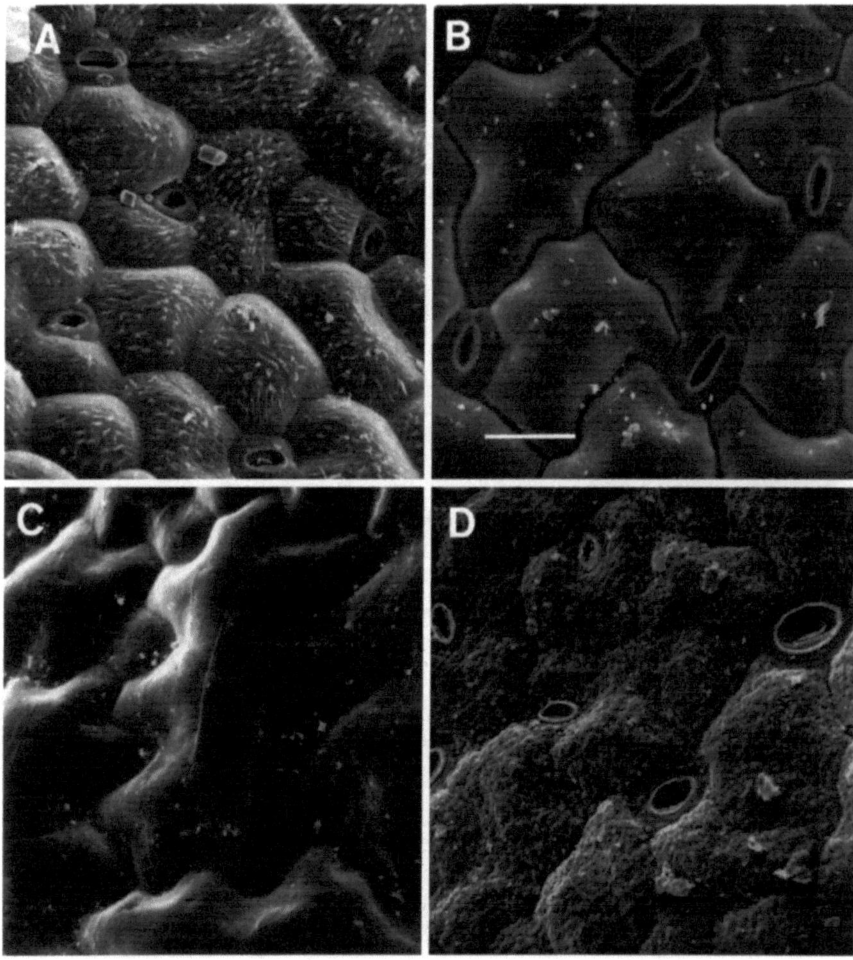

Fig. 3.18A–D. Adaxial leaf surfaces of desert herbs. **A** *Nemacladus rubescens*. **B** *Camissonia claviformis*. **C** *Pectis papposa*, a C_4 species. **D** *Anulocaulis annulata*. *Bar* 25 μm

ral common C_4 dicotyledonous annuals, such as *Aerva javanica* and species of *Amaranthus*, *Tribulus*, and many Chenopodiaceae (Sankhla et al. 1975). Most C_4 annuals have low A^{mes}/A value because chlorenchyma tends to be arranged around bundle sheaths, yet species of *Tribulus* have an adaxial layer of palisade mesophyll but what is described as an abaxial layer of water-storage cells (Volkens 1887).

Some common exceptions to the generalization about summer ephemerals are C_3 geophytic vines of Cucurbitaceae, which require warm soil temperature to initiate shoot growth, but are phreatophytic in that their roots tap deep moist soil reserves. Warm temperature is often required with heavy rainfall for germination of most woody perennials, hence these seedlings (except some

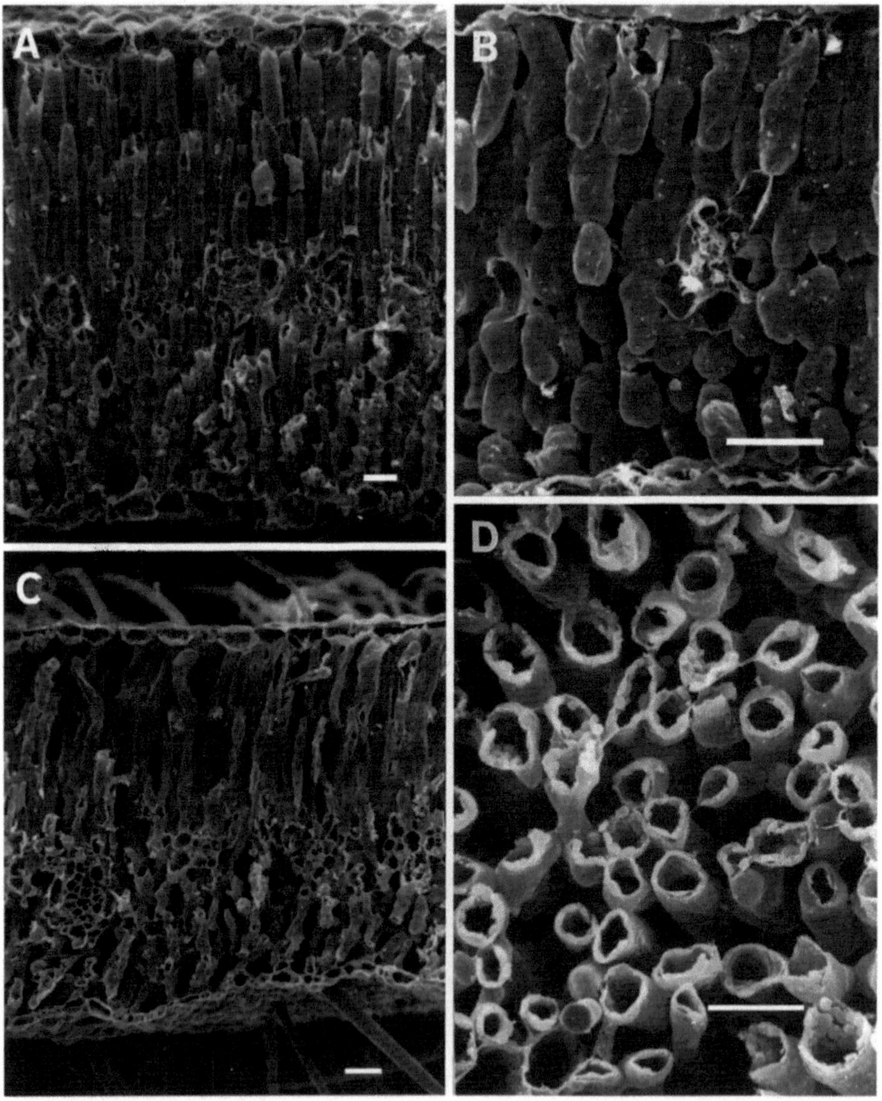

Fig. 3.19A–D. Leaves of desert annuals. **A** *Eriogonum brachypodum*. **B** *Lepidium lasiocarpum*. **C,D** *Camissonia brevipes*. **C** Transection. **D** Paradermal view of palisade mesophyll. *Bar* 25 µm

Chenopodiaceae) are arido-active C_3 herbs at the time. Desert cucurbits have amphistomatic leaves but bifacial mesophyll (Fig. 3.20A; Shields 1951a, North 1992).

In North America, arido-active broad-leaved *Datura wrightii*, which may cooccur with *Cucurbita*, can produce an A^{mes}/A of 30, and its leaf design is similarly bifacial (Fig. 3.20B). High adaxial and abaxial stomatal densities per-

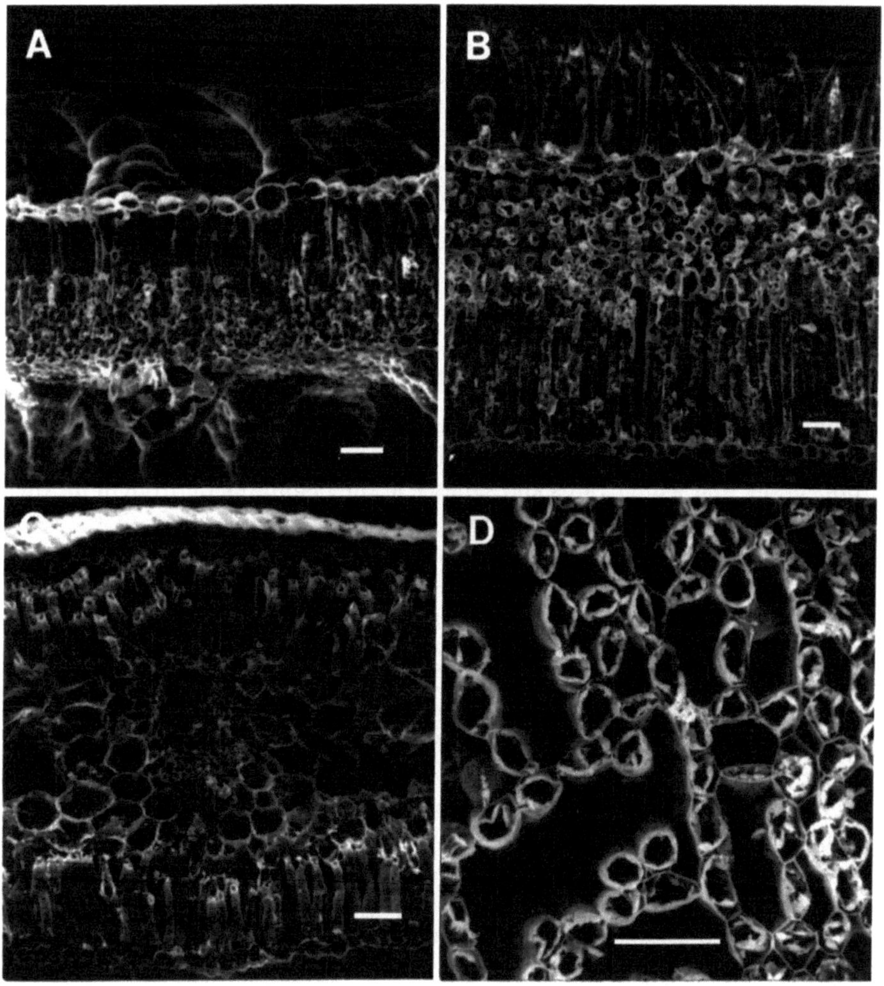

Fig. 3.20A–D. Leaves of North American herbaceous perennials. **A** *Cucurbita palmata*, with bifacial organization. **B** *Datura wrightii*, with bifacial organization. **C,D** *Hesperocallis undulata*. **C** Transection showing isolateral mesophyll. **D** Paradermal view of palisade mesophyll with well-developed substomatal chambers and intercellular air spaces. *Bar* **A–C** 50 μm; **D** 10 μm

mit this species to cool a leaf using extremely high rates of evaporation (Smith 1978).

Little physiological anatomy has been done on desert hemicryptophytes. *Ambrosia ambrosioides* and *Penstemon parryi* of wash habitats in the Sonoran Desert had relatively high photosynthetic rates of 20–25 μmol m^{-2} s^{-1} and 14–19 mmol m^{-2} s^{-1}, respectively (Knapp and Smith 1991). These California species, the very broad-leaved *Rumex hymenosepalus*, and the microphyllous *Stillingia paucidentata*, all have amphistomatic leaves with strongly to weakly isolateral mesophyll. In the Negev, e.g., *Reseda muricata, Heliotropium rotundifolium, Erodium glaucophyllum*, and *Haplophyllum tuberculatum*, have relatively thin cuticle and outer periclinal epidermal cell walls, superficial stomata on both sides of the leaf, and strongly to weakly isolateral mesophyll (Volkens 1887; Evenari 1938b). In the Sahara occurs *Abutilon muticum*, with soft, simple leaves to 50 mm wide, which are amphistomatic, having adaxial and abaxial stomatal densities of 488 mm^{-2} and mesophyll that is entirely composed of palisade cells (Stocker 1972).

3.4.3 Geophytes

Bulbous monocotyledons flower and leaf typically in spring and are exclusively C_3 species. In the Negev, *Bellevalia desertorum* and *B. eigii* form new leaves or roots at the first winter rains and senesce at the beginning of drought in April or May (Boeken 1990).

Leaves of *Hexacyrtis dickiana* have relatively thick cuticle and outer periclinal epidermal walls, hence slightly sunken stomata, thick-walled hypodermis, which may be three-layered along the keel, and chlorenchyma up to 16 layers thick (Baijnath 1988). *Zephyra elegans* of the Atacama and *Tulipa amplyophylla* of the Negev have no distinct palisade parenchyma (Evenari et al. 1982; Arroyo 1986), whereas *Hesperocallis undulatus* of the Sonoran Desert (Fig. 3.20C), *Dipcadi erythraeum* of the Sahara (Volkens 1887), and *Allium rothii* of the Negev have isolateral leaves (Evenari et al. 1982). All liliaceous desert species of Anatolia appear to possess amphistomatic leaves (Kasapligil 1961). In all monocotyledonous geophytes that have been observed, chlorenchyma is loose, having large intercellular air spaces (Fig. 3.20D).

Species of *Gynandrisis* and *Babiana hypogaea*, probably other African taxa, are geophytes that have coiled, narrow leaves. No study has determined whether this is simply a developmental oddity of these species or whether the design has an adaptive significance. Also lacking are anatomical descriptions of geophyllous species in southern Africa, which have unusual energy budgets (Eller and Grobbelaar 1982).

3.4.4 Desert Springs

Herbaceous perennials of freshwater desert springs are mostly obligate aquatics, that never occur elsewhere in the desert biome, disjuncts from aquatic habitats in moister climates. Because these are not xerophytes, little attention has been paid to them as desert organisms, inasmuch as they have anatomical adaptations of typical aquatic plants, such as thin cuticle and very loose chlorenchyma, often as aerenchyma. Desert species, like aquatics in other full-sun habitats of the world, are amphistomatic. An excellent example is the broad-leaved *Anemopsis californica* of the Mojave Desert.

3.5 Fruit Photosynthesis

In any terrestrial ecosystem, the majority of plant species form green reproductive organs; especially sepals, bracts, and pericarp may contribute to net carbon gain by the plant. Two potential avenues would be uptake of exogenous CO_2 and refixing endogenous CO_2 generated by high levels of tissue respiration within reproductive organs. As with leaves, CO_2 uptake requires the presence of stomata to create a reasonably high conductance for CO_2 diffusion into the organ and optimal development of palisade chlorenchyma with substantial A^{mes}/A.

Among desert taxa, fruit photosynthesis has rarely been investigated, e.g., in *Ambrosia deltoidea* (Szarek and Woodhouse 1977) and *Isomeris arborea* (Goldstein et al. 1991b). Substantial CO_2 uptake occurs in green fruit wings of *Atriplex canescens*, which exhibit Kranz anatomy and can have net photosynthetic rates during summer drought double that of water-stressed leaves (Rundel and Sharifi, unpublished data). It is likely that photosynthesis is important in other winged or inflated reproductive structures of desert taxa.

4 Photosynthetic Stems of Nonsucculent Plants

Photosynthesis by nonsucculent green stems contributes a major portion of annual carbon assimilation in numerous C_3 and some C_4 desert perennials. For example, in the Sonoran Desert, stem photosynthesis was responsible for 40% of net annual assimilation in *Cercidium floridum* (Adams and Strain 1969) and for 72% in *C. microphyllum*, mostly by twigs less than 6 mm in diameter (Szarek and Woodhouse 1978a, 1978b). Because precipitation in deserts is sporadic, in leafy species large seasonal and year-to-year fluctuations have been observed in the relative contribution of stems to net carbon assimilation. In shrubs with ephemeral microphylls, e.g., *Thamnosma montana* and *Salazaria mexicana*, annual stem assimilation accounted for 83% of net carbon assimilation (Comstock et al. 1988). Species of *Ephedra, Castela emoryi*, and *Koeberlinia spinosa*, in which leaf biomass is negligible, and aphyllous, fleshy-stemmed Chenopodiaceae, such as *Hammada scoparia*, can be expected to depend almost entirely on stem photosynthesis (Kappen et al. 1976; Schulze et al. 1976). Drought-deciduous perennials, such as *Gutierrezia sarothrae* (DePuit and Caldwell 1975), *Hymenoclea salsola* (Comstock and Ehleringer 1988), and other common chamaephytes (Comstock et al. 1988; Schmitt et al. 1993) may also depend substantially upon stem photosynthesis.

Green-stemmed microphyllous or aphyllous phanerophytes, belonging to many plant families, occur in all warm deserts (Table 4.1). Collectively these are a well-documented example of convergent evolution, including anatomical designs of stem tissues to function like leaf palisade for C_3 species (Böcher and Lyshede 1968, 1972; Went 1971; Evenari et al. 1971; Gibson 1983). Other green stems have evolved among C_4 species in a somewhat different way (James and Kyhos 1961; Fahn 1963; Carolin et al. 1975). For nitrogen-fixing perennial legumes, year-round stem photosynthesis appears to provide additional benefits to prevent nodule atrophy (Bossard and Rejmanek 1992; Nilsen 1995).

Both aphyllous green-stemmed species and leafy species, such as *H. salsola*, often prosper along wadis and runnels, where roots tap water-retaining sand and gravel layers (Zohary and Orshan 1956; Batanouny and Abdel Waheb 1973; Comstock et al. 1988), or are located in deep moist sand of hummocks. *Moringa peregrina* is one of several desert phreatophytic trees with substantial leaf canopies that utilize young-stem photosynthesis. Species that tap groundwater may exhibit relatively small diurnal fluctuations in shoot water potential and maintain high stem water potentials year-round (Szarek and Woodhouse

Table 4.1. Examples of nonsucculent green-stemmed woody plants from warm deserts (continent). Arrangement of cortical chlorenchyma is indicated as continuous (I), interrupted within internodes by major vascular bundles or collenchymatous strands (II), lining vertical grooves (III), or C_4 (see text), if known

Acanthaceae

Justicia californica (NAmer) II
J. hians (NAmer)
Monechma inarma (sAfr)

Amaranthaceae

Hermbstaedtia glauca (sAfr) C4

Apiaceae

Asteriscium chilense (SAmer)
Domeykoa oppositifolia (SAmer)
Eremocharis fruticosa (SAmer)
Gymnophyton foliosum (SAmer) II
Pituranthos aphylla (sAfr) II
P. scoparius (nAfr)
P. tortuosa (nAfr) II
P. triradiata (swAsia)

Apocynaceae

Skytanthus acutus (SAmer)

Asclepiadaceae

Asclepias albicans (NAmer) I
A. subulata (NAmer) I
Ectadium virgatum (sAfr)
Leptadenia pyrotechnica (nAfr-swAsia) I
Periploca aphylla (swAsia)

Asparagaceae

Asparagus aphyllus (nAfr)
A. spinosus (swAfr)
A. stipularis (nAfr-swAsia)

Asteraceae

Baccharis sarothroides (NAmer) II
Bebbia juncea (NAmer) II
Chloracantha spinosus (NAmer) II
Gutierrezia sarothrae (NAmer) II
Hymenoclea salsola (NAmer) II
Psila retamoides (SAmer) II
P. spartioides (SAmer) II
Rhantherium adpressum (nAfr-swAsia)
Stephanomeria pauciflora (NAmer) II

Brassicaceae

Oudneya africana (nAfr)
Vella mairei (nAfr)
Zilla macroptera (nAfr) I
Z. spinosa (nAfr-swAsia) I

Capparaceae

Cadaba aphylla (sAfr)
Capparis decidua (nAfr-swAsia) I

Celastraceae

Acanthothamnus aphyllus (NAmer) III
Canotia holacantha (NAmer) III
C. wendtii (NAmer) III

Chenopodiaceae

Haloxylon persicum (nAfr-swAsia) C_4
H. salicornicum (swAsia) C_4
Hammada scoparia (=*Anabasis articulata; nAfr-swAsia*) C_4
Maireana aphylla (Aust) C_4
Traganum nudatum (swAsia)

Cistaceae

Helianthemum kahiricum (nAfr)
H. scoparium (NAmer)

Crossosomataceae

Glossopetalon nevadensis (NAmer) I

Cucurbitaceae

Acanthosicyos horridus (sAfr)

Ephedraceae II (shallow furrows)

Ephedra alata (nAfr-swAsia)
E. alte (nAfr)
E. americana (SAmer)
E. andina (SAmer)
E. aphylla (swAsia)
E. aspera (NAmer)
E. breana (SAmer)
E. californica (NAmer)
E. campylopoda (sAfr)
E. fasciculata (NAmer)
E. foliata (swAsia)
E. funerea (NAmer)
E. nevadensis (NAmer)
E. trifruca (NAmer)
E. viridis (NAmer)

Fabaceae

Adesmia melanocaulos (SAmer)
A. microphylla (SAmer)
Alhagi pseudalhagi (nAfr-swAsia) I
Astragalus forskalii (nAfr) II
A. spinosus (nAfr)
Caesalpinia falcaria (SAmer) I
C. virgata (= Hoffmannseggia microphylla; NAmer) I
Cercidium australe (SAmer)
C. floridum (NAmer) I

Table 4.1 (*contd.*)

C. microphyllum (NAmer) I
Crotalaria aegyptiaca (nAfr)
Lotus rigidus (NAmer) II
Prosopis kuntzei (SAmer)
P. sericanth (SAmer)
Psorothamnus schottii (NAmer) II
P. spinosus (NAmer) II
Ramorinoa girolae (SAmer) II
Retama raetam (= *Lygos retama*; nAfr-swAsia) III
Senna aphylla (SAmer) II
S. armata (NAmer) II
Taverniera spartea (swAsia)
Goodeniaceae
Leschenaultia divaricata (Aust)
Koeberliniaceae
Koeberlinia spinosa (NAmer) I
Krameriaceae
Krameria grayi (NAmer) I
Lamiaceae
Salazaria mexicana (NAmer) II (collenchyma)
Loganiaceae
Logania sp. (Aust)
Malesherbiaceae
Malesherbia fasciculata (SAmer)
Malpighiaceae
Tricomeria usillo (SAmer)
Menispermaceae
Cocculus balfourii (swAsia)
Moringaceae
Moringa peregrina (nAfr-swAsia)
Oleaceae
Menodora spinescens (NAmer) I
Plumbaginaceae
Plumbago aphylla (SAfr)
Polygalaceae
Bredemeyera colletioides (SAmer)
Nylandtia spinosa (sAfr)
Polygala acanthoclada (NAmer)
Polygonaceae
Calligonum arich (nAfr) C_4
C. comosum (nAfr-swAsia) C_4
C. crinitum (swAsia) C_4
C. polygonoides (swAsia) C_4
C. tetrapterum (swAsia) C_4
Resedaceae
Ochradenus baccatus (nAfr-swAsia) II
Randonia africana (nAfr)
Rhamnaceae
Ziziphus obtusifolia var. *canescens* (NAmer) II
Rutaceae
Thamnosma montana (NAmer) I
T. trifoliata (NAmer)
Scrophulariaceae
Monttea aphylla (SAmer) I
Simaroubaceae
Castela emoryi (NAmer) I
C. polyandra (NAmer) I
C. stewartii (NAmer)
Solanaceae
Fabiana denudata (SAmer)
Verbenaceae
Diostea juncea (SAmer)
Junellia glavea (SAmer)
Neosparton darwinii (SAmer) III
N. ephedroides (SAmer) III
Zygophyllaceae
Bulnesia retama (SAmer) I
Sisyndite spartea (sAfr) I

1978b; Nilsen and Sharifi 1994; Smith et al. 1995). In contrast, those not utilizing groundwater may experience diurnal shoot water potential fluctuations of −3 MPa and have to survive shoot water potentials below −8 MPa (Kappen et al. 1975, 1976).

Year-round stem photosynthesis is generally associated with substantial stem transpiration (Wood 1934; Stocker 1971, 1974a, 1974b). This accounts for abundance of stem photosynthesis in wadis and runnels, and belies a common contention of some workers that these species have little water loss because they lack leaves.

A nonsucculent green stem may be interpreted as being adaptive if it meets any of the following criteria: (1) produces positive net carbon gain over the course of the year or for the lifetime of the structure; (2) permits survival or growth of a plant longer into the dry season by extending carbon uptake after leaves senesce; and (3) reclaims internal CO_2 (generated from respiration), especially when external CO_2 is not available to chlorophyllous cells during illuminated periods.

Classical examples of green-stemmed phanerophytes undoubtedly meet all three criteria. In comparison with leaves on the same plant, some green stems have a higher temperature optimum for photosynthesis (DePuit and Caldwell 1975) or better WUE (Comstock et al. 1988). Such stems possess features that are adaptive for improved carbon assimilation during summer months. However, in a Californian study on stem photosynthesis in *Senna armata* and *Caesalpinia virgata*, stems showed no spring to summer shift in thermal optima for photosynthesis, although there was some evidence for increased WUE during summer (Nilsen and Sharifi 1994).

4.1 Functional Morphology

4.1.1 Energy Budget Considerations

Daily energy budget of a fixed, cylindrical green stem differs from that of a comparatively thin, flat leaf [Eq. (2.1)]. Direct solar irradiation is maximally incident only on one half of each cylinder, possibly striking the entire cylinder on a given day but with a short arc perpendicular to the sun's rays at a given moment (Fig. 4.1). In particular, scattered skylight and reflected solar irradiation are major components of stem energy balance, but they vary markedly throughout the daytime and seasonally. PPFD is absorbed on the intercepting surface and probably not transmitted to the opposite chlorenchyma through central wood tissues, as it is through thinner leaves; consequently, amount of PPFD absorbed by the shaded surface may be smaller than that needed for light saturation, probably resulting in different rates of gas exchange on opposite surfaces.

Temperature of a narrow cylindrical stem (1–2 mm) remains close to ambient temperature even without transpirational cooling, and narrow stems have a thin boundary layer [less than 0.3 mm using Eq. (3.8)] and little heat-storage capacity. A green stem 5 mm in diameter still has a thin boundary layer (to 0.4 mm; Fig. 4.2) but a greater heat-storage capacity. Maximum stem temperature can be reduced by increasing reflectance of IR (Chap. 2), but this would reduce PPFD

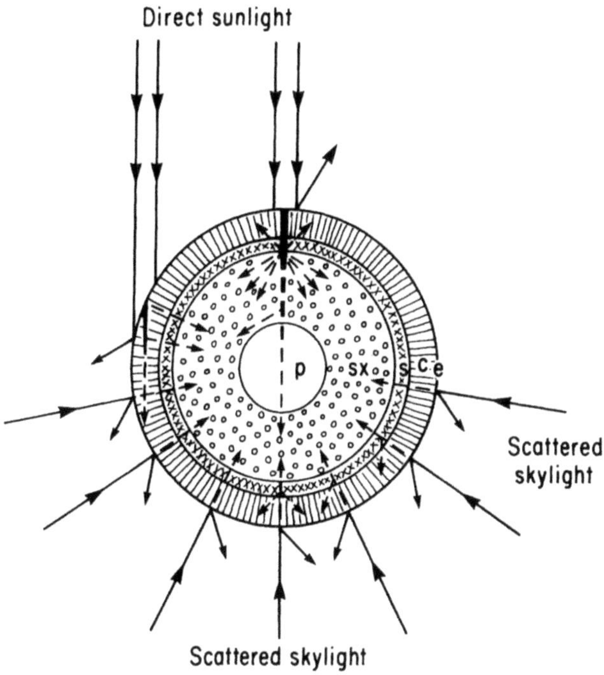

Fig. 4.1. A model for PPFD transmitted through a green stem that is several millimeters in thickness, assuming that the sun is perpendicular to the surface. Half of the cylinder instantaneously receives direct sunlight and the other half only scattered skylight. A portion of intercepted light will be reflected from the epidermis (*e*), including cuticle, cell walls, and trichomes covering the stem. PPFD is absorbed in palisade chlorenchyma (*c*) and attenuates (*dashed arrows*) while passing through deeper stem tissues. Some tissues, such as sclerenchyma in the phloem or inner cortex (*s*), secondary xylem (*sx*), and pith with secondary walls (*p*) probably reflect and refract light. PPFD intensity on the shaded half will be substantially lower, hence shaded chlorenchyma probably is not light-saturated

and thereby lower photosynthetic rate, as has been observed in water-stressed leaves of *Encelia farinosa* (Ehleringer and Björkman 1978b).

Presence of many leaves on shoots with green stems, or densely arranged stems in a canopy, would substantially reduce PPFD and infrared absorption by the average stem. Thus aphylly (apophylly of Böcher and Lyshede 1968, 1972), produced by extreme reduction of leaf size or shortness of leaf duration, represents an important adaptation to minimize shading of stem surfaces and to maximize PPFD absorption by individual stems.

4.1.2 Canopy Architecture and Phenology

Devising a formal classification for types of green-stemmed perennials has been unsuccessful because variations in shoot architecture are too numerous, rang-

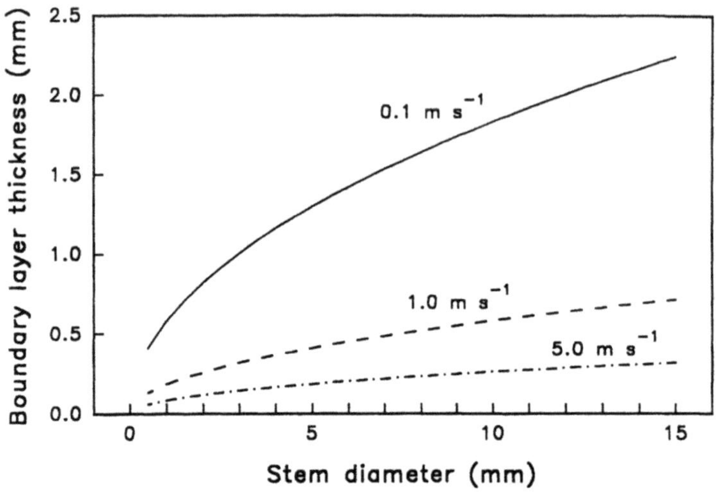

Fig. 4.2. Predicted boundary layer thickness (mm) for cylindrical stems of different diameters for calm (0.1 m s^{-1}), a gentle breeze (1.0 m s^{-1}), and a steady wind (5.0 m s^{-1})

ing from rigid to flexible, multi-year to one season, nonsucculent to fleshy or succulent, and leafy to essentially leafless.

One extreme in external canopy architecture is a "crucifixion thorn", a name applied to species with a spinose, three-dimensional canopy of evenly tapered, rigid green stems, as typified by *Koeberlinia spinosa* of North America (Fig. 4.3A). Most crucifixion thorns have sylleptic growth with branching that mimics the phyllotactic pattern because a straight branch arises from each axillary bud, e.g., helically alternate in *K. spinosa* (Fig. 1.6A), *Castela emoryi, Menodora spinescens, Psorothamnus spinosus,* and *Acanthosicyos horridus,* or opposite decussate in *Salazaria mexicana.* Some species resemble crucifixion thorns but do not have sylleptic branching, thereby lacking geometric regularity, and there are species having geniculate (zigzag) or divaricate sympodia, e.g., *Capparis decidua, Bredemeyera colletioides, Pituranthos tortuosa, P. triradiata, Eremocharis fruticosa,* and *Gymnophyton foliosum.*

At another extreme are species described as brooms or switch-branches that have long, slender, somewhat flexible, seldom branched, nearly parallel and fastigiate to ascending (virgate) green stems. Examples are *Bulnesia retama, Retama retaem, Ephedra* spp. (Fig. 4.3B), *Psila retamoides,* and *Neosparton ephedroides.* These tend to have proleptic growth and may have condensed branching with a shoot cluster at a node, although *Sisyndite spartea* is sylleptic. Species of *Calligonum* (C_4) and *Ephedra* have highly reduced ephemeral leaves that become nonfunctional when the shoot ceases vegetative elongation, whereas *R. retaem* is an example of a species that has leaves persisting for one or more months during the growing season. Low brooms, such as *Thamnosma montana, Senna armata,* and *S. aphylla,* form sympodial green shoots. In *S. armata,* a shoot terminates as a determinant inflorescence, dying back, so that in

Fig. 4.3A,B. A *Koeberlinia spinosa*, an example of the crucifixion thorn design. B *Ephedra nevadensis*, an example of broom design

the following year new growth arises from its most distal surviving axillary buds.

Neither orientation nor reflectance of photosynthetic cylindrical stems has been analyzed using appropriate computer simulations and direct measurements to determine any adaptive significance for crucifixion thorns and brooms. Based on analyses of flattened leaves (Chap. 2), canopy designs, especially fixed vertical orientation types, most likely represent strategies to maximize PPFD interception and possibly to reduce stem heating during summer, and limit photoinhibition (Ehleringer and Cooper 1992; Nilsen and Sharifi 1994). Dense trichomes occur on photosynthetic stems of certain species, notably the legumes *Psorothamnus spinosus* (Fig. 4.4A) and year-old axes of *Senna armata* (Fig. 4.4B), the mint *Salazaria mexicana* (Fig. 4.5F), and *Justicia californica* (Fig. 4.5E), suggesting a strategy to reflect IR, but the off-green colors of many species are attributable instead to thick cuticle. On perennials with very hairy, white mature stems, e.g., *Rhanterium epapposum*, young stems lack vestiture, presumably an adaptation that favors higher PPFD interception for photosynthesis.

Long, erect, flexible, monopodial green stems, often described as whiplike or wandlike, e.g., *Asclepias subulata* and *Leptadenia pyrotechnica*, develop thicker upper stems than typical broom plants and thereby exhibit a small degree of succulence. More typically classified as succulents are euphorbs, e.g.,

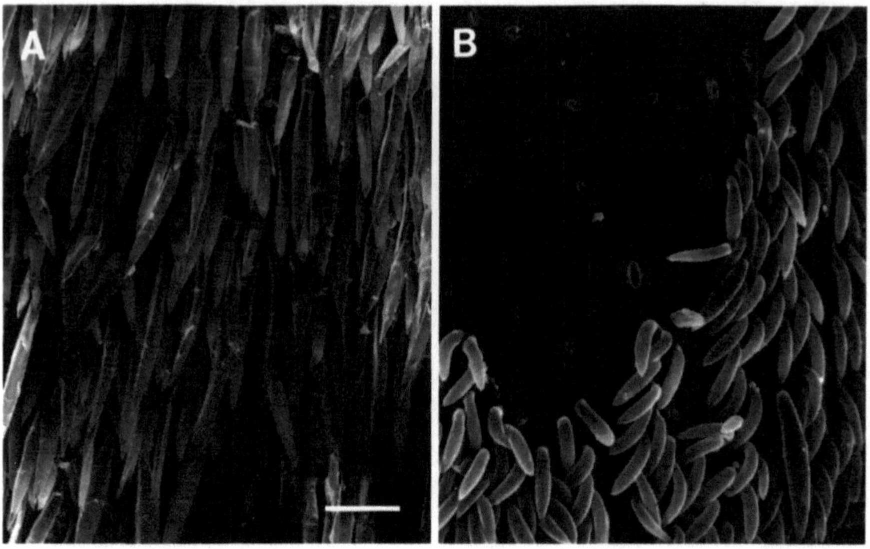

Fig. 4.4A,B. Reflective trichomes on green stems. A *Psorothamnus spinosus*. B *Senna armata*, showing zone where trichomes arise on a stem below current season growth. *Bar* 100 μm

desert species *Pedilanthus macrocarpus, Euphorbia antisyphilitica,* and *E. caducifolia.*

The above subtypes have stems that remain green for 2 or more years, and have leafless stems for most months of the year. Very long-lived are palo verdes of the genus *Cercidium*, which possess green main branches and trunks. *Cercidium floridum* has shoots that most closely resemble other leguminous brooms, whereas *C. microphyllum* is rigid, spinose, and somewhat divariate, with greater similarity to crucifixion thorns, although its leaves mostly arise from short shoots (Cannon 1908).

Substantial stem photosynthesis can occur on any newly formed organ that possesses chlorenchymatous outer cortex, especially with palisade-like cells, but, most importantly, has numerous stomata. This design is common in chamaephytes, e.g., *Farsetia aegyptiaca,* which has high stomatal density and three layers of palisade cells, and many Asteraceae. In such cases, stem photosynthesis remains positive following leaf senescence, allowing prolonged photosynthetic activity during drought months (DePuit and Caldwell 1975; Comstock and Ehleringer 1988). Young green stems that lack stomata probably only refix internal CO_2 and show negative net photosynthesis; these cases are termed corticular photosynthesis (Nilsen 1995).

Occurrence and importance of stem photosynthesis of shorter-lived shoots probably has been greatly underestimated. For example, in western North America, *Petalonyx thurberi* is a hemispherical chamaephyte along runnels and wadis with hundreds of green terminal stems that are photosynthetically active even during midsummer drought and probably are the reason this plant flow-

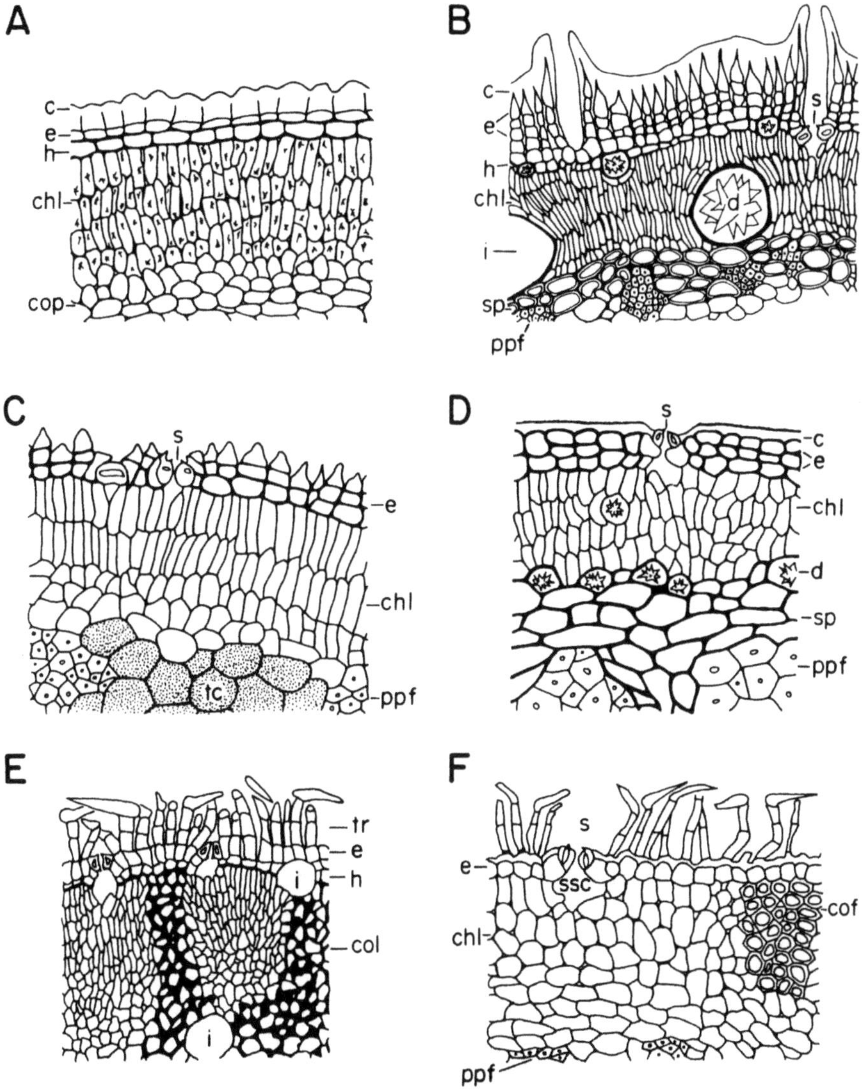

Fig. 4.5A–F. Transections of outer photosynthetic stems from North America. **A** *Thamnosma montana*. **B** *Castela emoryi*. **C** *Caesalpinia virgata*. **D** *Asclepias subulata*. **E** *Justicia californica*. **F** *Salazaria mexicana*. Key: cuticle *c*, chlorenchyma *chl*, cortical fibers *cof*, cortical parenchyma *cop*, druse *d*, epidermis *e*, hypodermis *h*, idioblast *i*, primary phloem fiber *ppf*, stoma *s*, sclerified parenchyma *sp*, substomatal chamber *ssc*, trichomes *t*. (Gibson 1983)

ers and sets fruit during drought months, later than sympatric shrubs. Young and relatively old stems of chamaephytic *Fagonia* are green. In similar habitats the hemicryptophyte *Stillingia paucidentata* forms a hemispherical canopy of photosynthetic stems with extremely narrow microphylls, all shoots that disappear during drought. On nearby dunes and sandy soils the tall, arido-active annual *Dicoria canescens* displays green stripes on stem beneath the thick layer of trichomes that reflect IR (Fig. 2.11C).

4.2 Anatomy of Woody Plants

Anatomical convergence is clearly evident in tissue and cellular features of these green-stemmed woody perennials. Characteristic features include: (1) long delay in formation of phellogen and periderm; (2) thick development of cuticle; (3) occurrence of numerous stomata; (4) concealment of stomata in pits, valleys, or furrows; (5) development of palisade-like cells in outer cortex; and (6) occurrence of sclerenchyma adjacent to palisade cortex.

4.2.1 Epidermis

4.2.1.1 Delay in Periderm Formation

Many desert phanerophytes have visible amounts of chlorophyll in woody stems (Cannon 1908), but bark is a poor pathway for gas diffusion, because gas diffusion through suberin is extremely slow. To maintain positive net uptake of CO_2, a stem must possess an epidermis with stomata that retain their function for months or even years. Therefore, periderm formation must be suppressed to permit guard cells to control gas exchange.

Few studies have carefully documented when phellogen formation begins in these aphyllous, green-stemmed species. Typically, phellogen arises superficially in epidermis or hypodermis, infrequently from chlorenchyma. Phellogen is initiated in short, widely separated arcs, so that such stems have patches or strips of bark interrupting green sectors. Crucifixion thorns and brooms have stems that can be photosynthetic for several years, so that periderm appears only on thick stems, which are also the ones susceptible to heating above air temperature. On leafy chamaephytes like found in Asteraceae, periderm forms during the first year.

4.2.1.2 Composition

Most species have a uniseriate epidermis, consisting of groundmass cells often having slightly convex outer walls (Fig. 4.5A; Gibson 1983). Multiple epidermis, formed by several synchronous periclinal divisions of epidermal cells, has been described in *Castela* spp. (Fig. 4.5B), *Canotia* spp., *Bulnesia retama*, *Caesalpinia virgata* (Fig. 4.5C), *Ramorinoa girolae*, and *Bredemeyera colletioides* from American deserts (Böcher and Lyshede 1968; Gibson 1979, 1983). In these cases, very thick cuticle isolates the cell clusters. Other examples of multiple epidermis are found among caesalpinoid legumes, *Leptadenia pyrotechnica* (Stocker 1971), *Asclepias subulata* (Fig. 4.5D; Gibson 1983), and *Pituranthos tortuosa* (Volkens 1887). Development of multiple epidermis is considered to be an adaptation to accommodate an increase in stem diameter without affecting epidermal integrity or disrupting guard cell function (Böcher and Lyshede 1972).

Paradermal sections of green stems often show that each protodermal cell functions as an initial to form a polygonal array of groundmass epidermal cells (Böcher and Lyshede 1972). This results in increases in stem girth in a manner similar to that observed in succulents (Gibson and Horak 1978).

4.2.1.3 Surface Coverings

Mature green stems are superb examples of plant organs with highly developed cuticle. Desert and semidesert species have been subjects of pertinent histochemical and ultrastructural studies and adaptational analyses (Böcher and Lyshede 1968, 1972; Lyshede 1977a,b,c,d, 1978, 1979); however, accompanying measurements of g_{wv} have not been made to determine which designs are most effective, or how they compare with conductance of leaves on the same or different species or of succulent stems. While allowing CO_2 uptake, epidermis should have extremely low g^c to limit transpirational losses when stomata are closed, but still permit high PPFD to reach the underlying chloroplasts.

Waterproofing waxes are characteristically thick, exceeding 5 μm, and thickening may occur either as primary cuticle (cuticularized layer), secondary cuticle (cutinized layer complexed with cell wall polysaccharides), or both. One famous example is *Monttea aphylla* (Fig. 4.6A), in which primary cuticle may

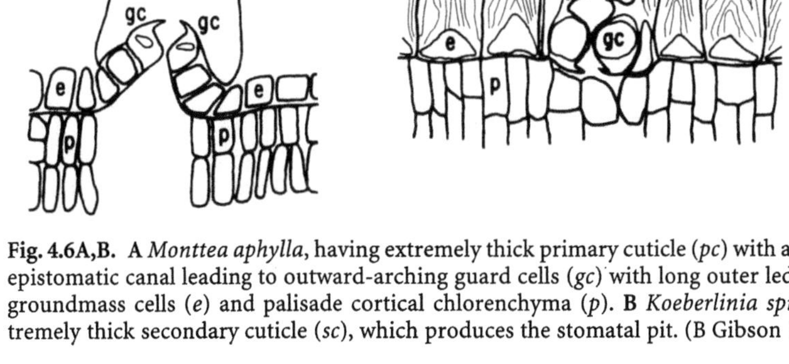

Fig. 4.6A,B. A *Monttea aphylla*, having extremely thick primary cuticle (*pc*) with a relatively wide epistomatic canal leading to outward-arching guard cells (*gc*) with long outer ledges; epidermal groundmass cells (*e*) and palisade cortical chlorenchyma (*p*). B *Koeberlinia spinosa*, have extremely thick secondary cuticle (*sc*), which produces the stomatal pit. (B Gibson 1979)

be 180 μm in thickness (Böcher and Lyshede 1968). *Koeberlinia spinosa* (Fig. 4.6B) has a cutinized outer wall to 60 μm in thickness (Gibson 1979). Cuticular flanges are conspicuous in *Senna aphylla, Psorothamnus spinosus,* and *Salazaria mexicana,* to name a few. Inner periclinal walls are cutinized in *Ochradenus baccatus*. This feature would lower g_{wv}.

The majority of photosynthetic stems are glabrous or glabrescent (Figs. 4.7A–D), thereby lacking structures that shade chlorenchyma. In *Justicia californica* and *Salazaria mexicana,* L-shaped uniseriate nonglandular trichomes develop from the groundmass cells to form a tight mat (Fig. 4.5E–F). Each trichome has a heavily cutinized terminal cell that is oriented parallel to

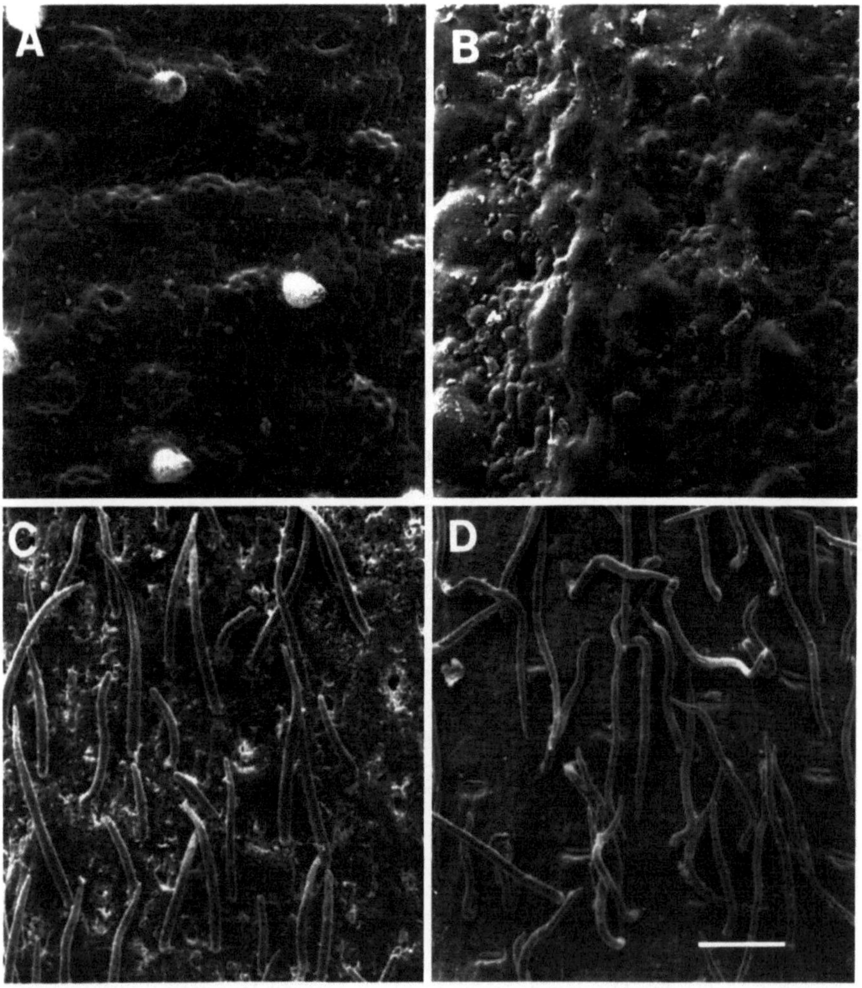

Fig. 4.7A–D. Surfaces of green stems from North American deserts. **A** *Koeberlinia spinosa*. **B** *Ephedra trifurca*. **C** *Caesalpinia virgata*. **D** *Krameria grayi*. *Bar* 300 μm

the stem surface (Gibson 1983), and the stalk cells also have cutinized walls. *Farsetia aegyptiaca* has reflective, T-shaped trichomes that cover the epidermis. In *Petalonyx thurberi*, family-specific barbed trichomes partially hide green stems until late summer, when epidermis and hypodermis form a whitish, highly reflective, shell-like stem coating. Many legumes have nonglandular unicellular trichomes, bent at right angles to the base and pointing toward the apex (Fig. 4.4).

4.2.1.4 Stomata

Stems typically have moderate stomatal densities, ranging from 50 to 190 mm^{-2}, which are low values as compared with either abaxial or adaxial stomatal densities on leaves of the same plant, but the guard cells are slightly longer, so therefore have larger pores when fully open. Ridges formed over vascular bundles or other nonphotosynthetic tissue lack stomata. Orientation of stomatal pores is longitudinal for most species though transverse in some legumes and *Krameria grayi* (Fig. 4.7D). No functional differences should result from having different orientations, all other parameters being equal.

Proliferation of protoderm as multiple epidermis or formation of thick primary and secondary cuticle results in stomata becoming sunken, positioned beneath the outer stem surface at the same distance as the eccentric growth. The majority of green stems with uniseriate epidermis have guard cells located in the center of shallow pits 5–20 μm in depth. Exceptions are *Koeberlinia spinosa*, sunken to a depth of 40 μm (Fig. 4.6B; Gibson 1979, 1983), and *Monttea aphylla* at depths exceeding 100 μm (Fig. 4.6A; Böcher and Lyshede 1968). Species with multiple epidermis characteristically have more deeply sunken stomata, commonly exceeding 20 μm and greatest in *Castela emoryi* (Fig. 4.5B) and *Bredemeyera colletioides*, which have stomatal chimneys or stomatal pitchers, to a depth of 150 μm (Bessey 1904; Böcher and Lyshede 1968; Wilkinson 1979; Gibson 1983). Stems with longitudinal furrows, e.g., *Canotia holacantha* and *Retama retaem*, have stomates up to 80 μm from the projected cylindrical outline. Stomatal crypts, having a number of stomata per crypt, have been observed in *Menodora spinescens*, and *Glossopetalon nevadensis* has one stoma at the base of each crypt (Gibson 1983).

As discussed in Chapter 3, sunken stomata cannot be viewed as adaptations to reduce transpiration, unless suprastomatal cavities are very deep, e.g., stomatal pitchers in *Castela emoryi* or thick cuticle of *Monttea aphylla* and *Koeberlinia spinosa*, where estimated suprastomatal conductances were less than 7 mm $s^{-}1$. Predictions are that when stems are fully hydrated and stomata are wide open, only very long suprastomatal pathways would decrease transpiration as well as CO_2 uptake. Reducing transpiration by this method would probably lower carbon uptake, but presumably at a higher WUE. Deeply recessed guard cells hypothetically would experience a different microclimate than would superficial stomata, e.g., PPFD, humidity, or temperature, producing an adaptation to permit stomata to open on days when hot, dry conditions would otherwise shut down stem photosynthesis (Schulze et al. 1974, 1975a, 1975b; Lange et al. 1975a; Landsberg and Butler 1980; Lösch and Tenhunen 1981; Grantz 1990). Thick, cutinized epidermis most likely is the important adaptation, for protecting long-lived stem chlorenchyma from UV damage.

Stems having relatively thin cuticle, which is then coated with resin, have been described in *Fabiana* and *Psila* (Böcher and Lyshede 1972) and occur in *Hymenoclea salsola* by being produced from glandular trichomes. Guard cells in these species typically are raised and often form conspicuous outer ledges, presumably to avoid plugging by resin, although some blockage by resin still occurs. Highly elevated guard cells occur on resinous green stems of *Baccharis sarothroides* and other photosynthetic stems of Asteraceae.

4.2.2 Hypodermis

Hypodermis is present in less than one-quarter of the species that have been described and typically is weakly developed among warm-desert representatives, being more common in green-stemmed species of cool deserts and semi-arid vegetation types (Böcher and Lyshede 1972). Uniseriate, collenchymatous hypodermis occurs in *Psorothamnus spinosus, Justicia californica* (Fig. 4.5E), and *Senna armata,* which are stems that have dense trichome covers but no sunken stomata (actually raised in *J. californica*), as well as in glabrous stems having shallowly sunken stomates, e.g., *Crotalaria aegyptiaca* (Lyshede 1977a), *Prosopis kuntzei* (Böcher 1975), and *Thamnosma montana* (Fig. 4.5A; Gibson 1983), as well as *Castela emoryi* (Fig. 4.5B; Gibson 1983) and *Ramorinoa girolae,* which have deeply sunken stomata. Thick-walled hypodermis occurs in South American species of *Senna,* and in *Menodora spinescens* of cool desert and semidesert habitats, the hypodermal layer consists of very thick-walled gelatinous fibers, which are absent around the stomatal crypts (Gibson 1983).

Effects of hypodermis on stem physiology have not been quantified for any of these species, although a role in reducing transpiration has been suggested (Lyshede 1977a). To be investigated are its possible significance as a radiation filter affecting UV, PPFD, and IR, or how hypodermis may influence longevity of the epidermis during increases in stem age and diameter.

4.2.3 Chlorenchyma and Associated Sclerenchyma

Green-stemmed woody plants of warm desert communities characteristically have a continuous or interrupted chlorenchymatous band 60–175 μm thick and composed of palisade-like outer cortical cells with thin primary walls (Fig. 4.5A–F). The inner edge of chlorenchyma typically is clearly defined by an anatomical feature, such as crystalliferous cells, a starch sheath, tannin cells, ground tissue with secondary cell walls, sclerenchyma, occasionally endodermis, or a combination of these, depending on the synapomorphic feature of each genus. Especially in species with rigid stems, brachysclereids commonly differentiate between clusters of primary phloem fibers to complete a ring of lignified cells, separating chlorenchyma from sieve tubes of the vascular cylinder, but in the same way encasing functioning vascular tissues.

Among dicotyledons, number of cell layers commonly is three or four but varies from seven, as in *Canotia wendtii* and *Justicia californica* (Fig. 4.5E), to one, as in *Krameria grayi*. Often, layers are not rigidly defined, and inner ones or even certain of the species have nearly cuboidal or isodiametric rather than elongate chlorenchyma cells. Chlorenchyma cells commonly are radially stretched during increase in stem girth, accounting for measurable increases in chlorenchyma thickness from young to old green stems. Rarely, sclereids differentiate within palisade layers, as in *Capparis decidua* (Stocker 1971) and *Bulnesia retama* (Böcher and Lyshede 1968).

Many aphyllous stems have palisade chlorenchyma with conspicuous intercellular air spaces and very little touching of lateral walls (Fig. 4.8). Nonetheless, cortical chlorenchyma, which is initiated from stem ground meristem, appears to differ somewhat in its development from that of leaf mesophyll in that formation of intercellular air spaces may be more constrained, resulting in greater touching of stem chlorenchyma. This is probably due to less cell enlargement of groundmass epidermal cells, which in leaves causes maturing mesophyll cells to be pulled apart (Isebrands and Larson 1973). Substomatal chambers are well developed in stem chlorenchyma (Fig. 4.9), as they are in leaf mesophyll.

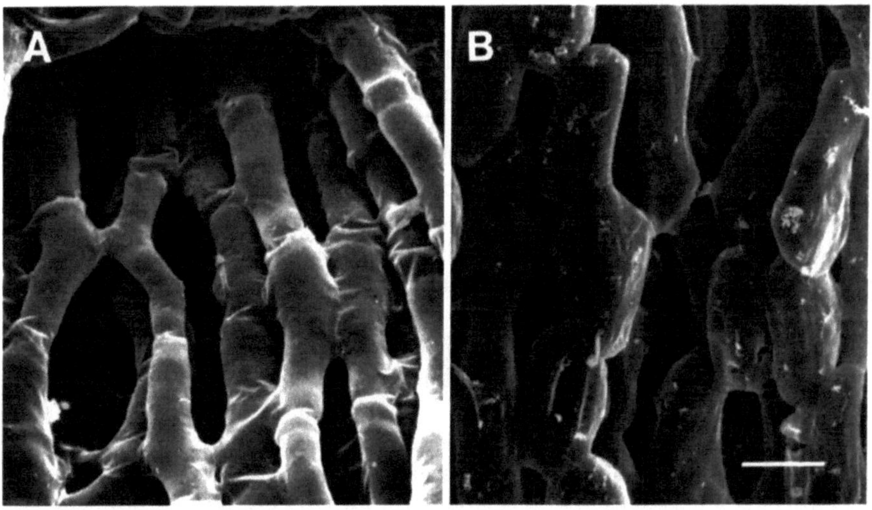

Fig. 4.8A,B. Cortical palisade chlorenchyma with well-developed intercellular air spaces and low degree of contact between cells. **A** *Monttea aphylla*. **B** *Thamnosma montana*. Bar 10 μm

Fig. 4.9. Paradermal view of palisade chlorenchyma in a stem of *Caesalpinia virgata*, showing conspicuous substomatal chamber. *Bar* 25 μm

Chlorenchymatous cortex is compact yet tends to have small but well-developed intercellular air spaces because anticlinal cell walls are connected usually less than one-third. In the literature, drawings of these stems rarely have shown these intercellular air spaces because thick sections were used, but SEM photomicrographs clearly illustrate these important diffusion pathways for CO_2 to cell walls (Fig. 4.8A,B). However, in stem chlorenchyma of a few species, e.g., *Koeberlinia spinosa*, less than 20% of anticlinal walls are free. Low exposure of cell wall occurs in stem chlorenchyma of *Balanites aegyptiaca*, which has numerous stomata, but probably signifies that photosynthetic rates of these stems are comparatively low.

Although various authors have interpreted small intercellular air spaces as a handicap to diffusion (e.g., Böcher and Lyshede 1972), applying Fick's first law for gas diffusion [Eq. (3.5)], length of the diffusion pathway, not size or shape of this comparatively large space for gas molecules, is the important parameter. Using this reasoning, one would predict that CO_2 diffusion from stomata to chlorenchyma cell walls is not slowed by having small air volume, because CO_2 conductance through stomata is much lower than gas phase conductance within the leaf unless the stomatal pore-to-chlorenchyma distance is greater than 0.4 mm. Water vapor conductance should not be affected by size or shape of intercellular air spaces, which are saturated with water vapor, because the principal sites of evaporation are noncutinized cell walls directly adjacent to the guard cells.

No quantitative data have been published on internal cell wall area of stem chlorenchyma as compared with leaf A^{mes}/A (Chap. 3) on the same plant. In many species, palisade mesophyll is thicker in leaves, cell density is higher, and cell touching is lower, so a higher ratio and therefore higher photosynthetic rate is predicted for leaves, where there would be more free cell wall for liquid-phase CO_2 diffusion. Nonetheless, in *Hymenoclea salsola*, a species in which stem chlorenchyma was found to be thicker, leaves still exhibited higher absolute

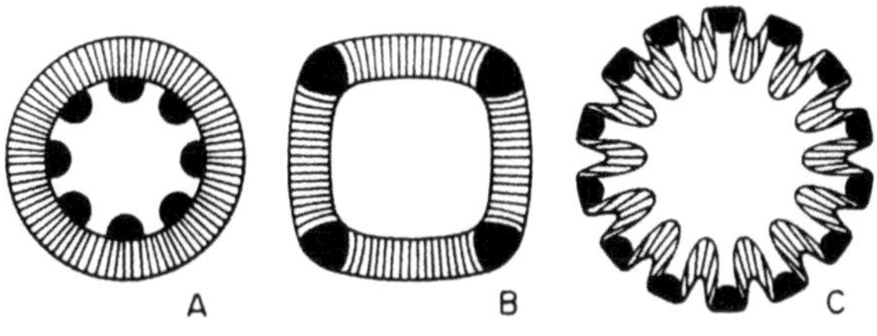

Fig. 4.10A–C. Three types of basic chlorenchyma designs in nonsucculent green stems of deserts. **A** Type 1 has continuous palisade layers in the outer cortex. **B** Type 2 has cortical palisade that is interrupted by strands of thick-walled fibers (*solid black*). **C** Type 3 has palisade chlorenchyma around furrows and strands of fibers within each stem ridge. (Gibson 1983)

rates of net photosynthesis, especially higher quantum yield efficiency, higher photosynthetic rate under saturated CO_2 concentration, and higher light saturation (Comstock and Ehleringer 1988). Conclusions drawn from *H. salsola* should be applicable to other Asteraceae like *Stephanomeria pauciflora, Bebbia juncea* (Schmitt et al. 1993), and species of *Chrysothamnus,* which have photosynthetic stems that persist after leaves have suffered extreme water stress or have been shed.

Three basic types of nonsucculent mature woody stems occur in warm deserts (Böcher and Lyshede 1972; Gibson 1983). In most genera of these old green stems, chlorenchyma of internodes occurs either as a continuous cylinder (Type 1; Fig. 4.10) or a cylinder interrupted by leaf-trace bundles with conducting xylem and phloem or panels of sclerenchyma (collenchyma is a few), continuous with support tissues of distal leaves (Type 2; Fig. 4.10). Presence or absence of leaf-trace bundles, or associated support cells within cortex of stem internodes is undoubtedly determined by the nature of the ancestral primary vasculature pattern, in that in some lineages leaf-trace bundles diverge from the vascular cylinder one or more nodes before entering the base of a leaf primordium, whereas for others the leaf-trace bundles diverge very close to a node toward which it diverges (Gibson 1983). Unlike leaves, a typical stem transection shows no sieve tubes located within or next to chlorenchyma, except where a bundle supplies a leaf, so that export of sucrose is for these species still to be explained but would have to involve cellular transport along a parenchyma cell pathway. Type 3 are stems with deep longitudinal furrows, having chlorenchyma concentrated around the furrows and fibers localized in the ridges (Fig. 4.10).

Mechanical engineering studies of these green stems would be instructional to explain how patterns of sclerenchyma, or collenchyma in the case of *Justicia californica* (Fig. 4.5E), contribute to the evolutionary design of this life form. Certainly stems have clusters of strong and flexible primary phloem fibers or cortical fibers arranged in parallel and uniformly around the periphery to resist tensile and shear stress components, because such fibers have high elastic moduli and yet produce bending stiffness without deformation or breaking (Niklas 1992). Especially broom-like designs would be predicted to produce high frequencies of vibrations that could lead to shearing of leaves, whereas branching, as in crucifixion thorns, may reduce the magnitude of shearing. In *Ephedra* and *Asclepias subulata*, flexible stems contain prominent clusters of thick-walled gelatinous and unlignified fibers, whereas in rigid stems, sclerenchyma is precocious in its development and heavily lignified.

Both Type 1 and Type 2 stems are found among crucifixion thorns and brooms, whereas Type 3, having furrowed stems (longitudinal grooves), contains species with ascending or erect branches and is known from only a few desert brooms, best expressed in *Canotia holacantha* (Gibson 1979), *Retama retaem* (Volkens 1887; Evenari 1938b; Evenari et al. 1982; Fahn 1990; Fahn and Cutler 1992), *Pituranthos tortuosa* (Evenari et al. 1982), *Neosparton ephedroides* (Böcher and Lyshede 1972), and *Plumbago aphylla*. In Type 3 stems, chlorenchyma around the furrow is isolated from chlorenchyma of the adjacent furrow by a cluster of fibers localized in each ridge. Each cluster of ridge fibers is related to and opposite a leaf-trace bundle.

The three basic designs of stem chlorenchyma may have physiological significance, although no studies have pursued this line of research. Type 1 parallels the design of many desert leaves, which have no prominent midvein or secondary veins, and therefore when light-saturated photosynthetic capacity of each unit area around the cylinder should be roughly the same. Type 2, as with leaves having well-developed veins and associated support tissues, will have both photosynthetic and astomatic, nonphotosynthetic sectors, and, for a given species, only 30–80% of stem surface will be used for carbon assimilation (for leafy Asteraceae, see Comstock et al. 1988). Type 3 stems form 20–40% greater chlorenchyma than Type 1 stems having the same diameter. Support tissues in Type 2 and 3 influence PPFD absorptance by and transmission through stem chlorenchyma, diminishing chlorophyll absorptance at some angles and possibly reflecting light back into chlorenchyma in others. Especially in Type 3 stems, where deep-seated chlorenchyma cannot receive direct illumination and trichomes or epidermal papillae occur only in furrows, the role of fibers in light interception needs to be explored. Schanderl (1935) speculated that subepidermal fiber bundles may function as avenues for deeper penetration of PPFD into chlorenchyma.

For typical crucifixion thorns and brooms, it remains unexplained why chlorenchyma terminates less than 250 μm from the surface. PPFD should be able to penetrate more deeply into stems before attenuating, and chlorophyll has been observed within pith of 1.5-mm stems of *Larrea tridentata* (Cannon 1908), throughout mesophyll of moderately thick leaves, or within succulent stems as deep as 3–4 mm from the surface. Internal CO_2 concentrations of cortical chlorenchyma appear to be much lower than of leaf mesophyll, which may be related to some inherent structural difference of these two tissues (Ehleringer et al. 1987; Comstock and Ehleringer 1988).

Early maturation of primary phloem fibers, differentiation of cortical or medullary sclerenchyma, and precocious secondary growth are common features of most green-stemmed desert species (Gibson 1983). Although hardness of these species may be an adaptation to reduce vertebrate herbivory, as well as a way to achieve optimal orientation for managing stem energy budgets, research should also determine if these designs prevent physiological problems in vessels or sieve tubes.

4.2.4 Leaf Anatomy

Leaves of green-stemmed phanerophytes closely fit the anatomical profile of typical desert shrubs (Chap. 3): amphistomatic microphylls (Table 3.2) with isolateral palisade mesophyll (Fig. 4.11A–E). Epidermis is uniseriate, typically with

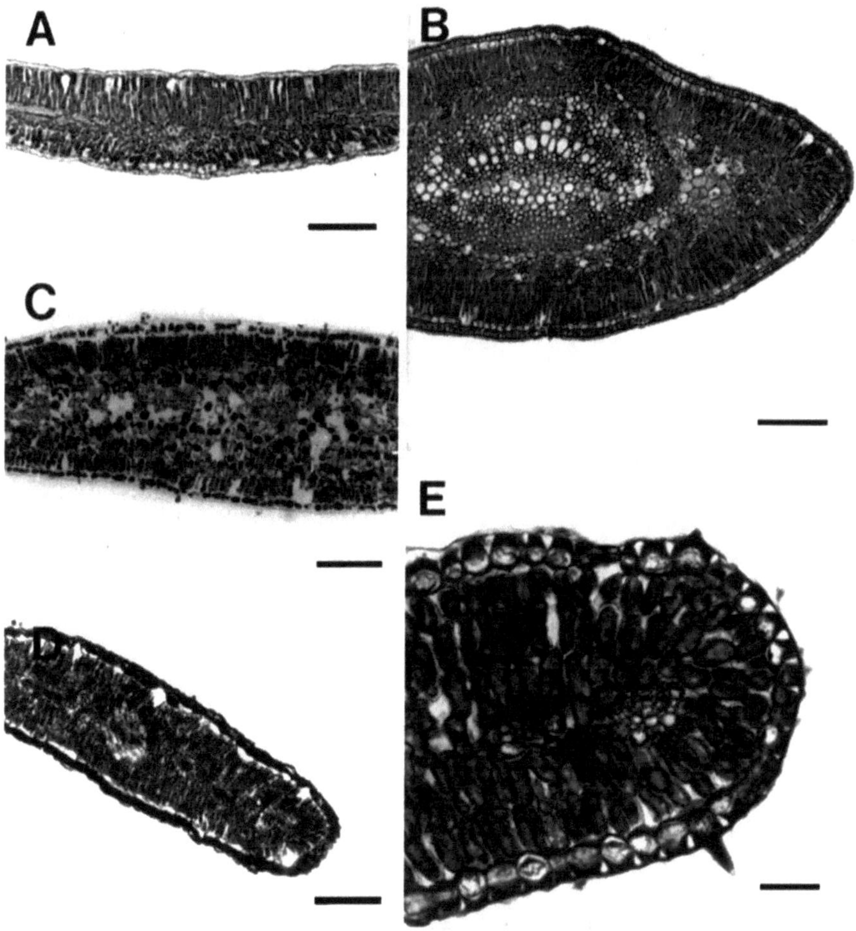

Fig. 4.11A–E. Transections of ephemeral leaves from green-stemmed shrubs of North American deserts. **A** *Thamnosma montana*. **B** *Asclepias subulata*. **C** *Cercidium microphyllum*. **D** *Caesalpinia virgata*. **E** *Menodora spinescens*. *Bar* A–D 100 μm; E 25 μm

superficial stomata or those sunken several micrometers; however, *Castela* has multiple epidermis and deeply sunken stomata, as on stems, and these ephemeral leaves have conspicuous sclerenchyma associated with vascular bundles. A few species with minute, vestigial leaves may have stomata only on one leaf surface, and in *Koeberlinia* stomata are deeply sunken (Gibson 1979).

Whereas on most photosynthetic stems the microphylls are shed as soon as the stem becomes sclerified or experiences slight water stress, the pinnately compound leaf of certain legumes may lose leaflets but retain a photosynthetic rachis. For warm deserts this is best illustrated by *Senna armata* (Fig. 4.12). This species forms leafy, flowering shoots in April and May, sheds its relatively thick, isolateral leaflets, but retains the flattened rachis, which has numerous stomata (>100 mm^{-2}) and some development of palisade parenchyma, thus extending leaf photosynthesis into early summer. Strong development of sclerenchyma within the rachis appears to block abscission until the following spring. Another example of a photosynthetic flattened rachis is the phreatophytic tree *Parkinsonia aculeata*, which originated in dry woodland but is cultivated and naturalized in deserts of the world (Scott 1935b).

Leaves have higher stomatal density but shorter stomatal pores than the green stems on which they are formed. In *Capparis decidua*, leaves had 40% more adaxial than abaxial stomata on leaves and significantly higher densities on leaves than on young to mature stems (Lekhak et al. 1983). Whereas orien-

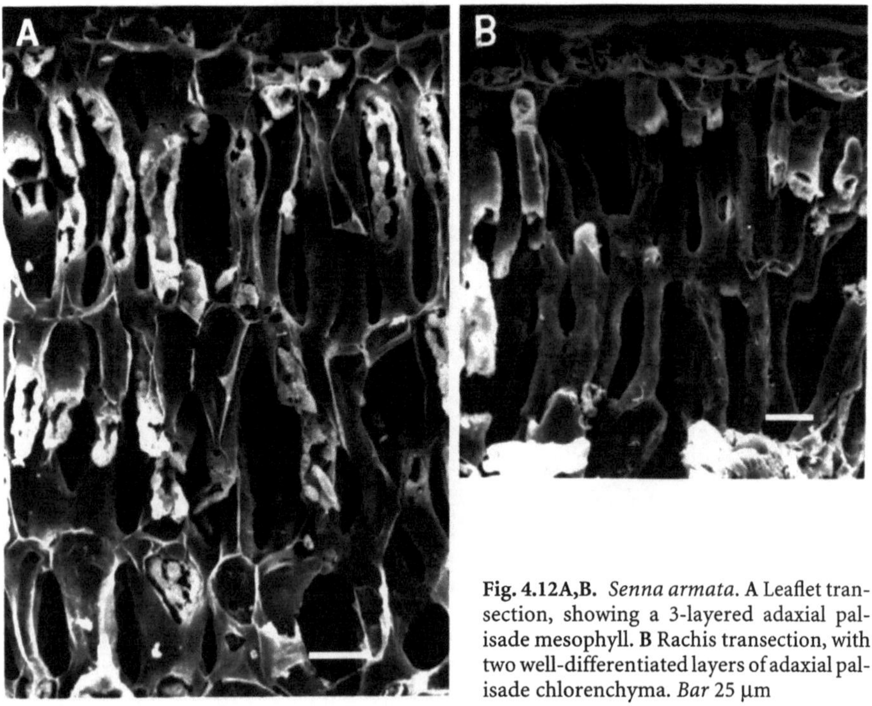

Fig. 4.12A,B. *Senna armata.* **A** Leaflet transection, showing a 3-layered adaxial palisade mesophyll. **B** Rachis transection, with two well-differentiated layers of adaxial palisade chlorenchyma. *Bar* 25 μm

tation of stomatal pores on stems is strictly either transverse or more commonly longitudinally, orientation on lamina is typically random. Longitudinal orientation characterizes stems and narrow leaves of *Stillingia paucidentata* of California and its green-stemmed relatives in the Monte (Böcher and Lyshede 1968).

Zilla spinosa, an Old World biennial or perennial with photosynthetic stems, has large basal leaves with groundmass epidermal cells having thin, undulating anticlinal cell walls and relatively low A^{mes}/A versus microphyllous cauline leaves with thicker, straight epidermal walls and higher A^{mes}/A due to the occurrence of thinner and longer palisade mesophyll cells.

4.3 Special Cases

4.3.1 Bark Photosynthesis of Cercidium

Cercidium, known collectively as palo verdes, is a genus of arborescent legumes from semiarid to arid habitats having prominent green stems. For desert species, depth of chlorophyll is at least 2 mm in stems of *Cercidium microphyllum* and up to 6 mm in *C. floridum* (Cannon 1908), and these species also differ from typical green-stemmed desert shrubs in producing leaves with microphyllous leaflets under favorable air temperatures and soil water supplies. Unlike typical bark photosynthesis of forest or woodland trees (Pearson and Lawrence 1958; Strain and Johnson 1963; Foote and Scheadle 1976; Han and Suzaki 1981), which results in no net carbon gain using exogenous CO_2 because few or no stomata are present in the gas diffusion pathway, in *Cercidium* numerous stomata persist even on the oldest trunk, and phellogen formation is thus suppressed.

Stem anatomy of *Cercidium* (Cannon 1908; Scott 1935b; Went 1971; Gibson 1983) is markedly different from that of aphyllous green-stemmed shrubs. *Cercidium* has a type of multiple epidermis formed by precocious anticlinal and periclinal divisions of a uniseriate epidermis, and these divisions continue for years as stem girth increases, yielding a bullate surface with the original stomata persisting in depressions or horizontal grooves. Youngest stems form a substantial cutinized surface 3–5 μm thick, and thick primary cuticle accumulates as stems age, most likely serving as a UV radiation screen as well as to reduce cuticular transpiration.

Cercidium has Type 1 chlorenchyma (Fig. 4.10), in stems of all sizes. Palisade-like chlorenchyma cells increase in volume to accommodate increases in stem diameter (Cannon 1908), mostly via anticlinal divisions (Gibson 1983). In *C. floridum*, chlorenchyma in large branches may be double in thickness as compared with young stems, and here multiseriate hypodermis, consisting of large, highly vacuolate parenchyma, occurs between outer chlorenchyma and multiple epidermis, except directly opposite guard cells (Scott 1935b). Experiments are required to determine whether water-storing hypodermis is an adaptation to maintain high chlorenchyma and epidermal water potentials for prolonging their functions. Much precipitation of calcium oxalate crystals occurs in cells on both tissues touching hypodermis. By year 2, chlorenchyma is isolated from translocating phloem by a ring of primary phloem fibers and sclerified parenchyma.

4.3.2 C_4 Chenopodiaceae and *Calligonum*

Numerous green-stemmed perennials with C_4 photosynthesis are especially common in either sand or salt deserts of Asia and northern Africa (Lyshede 1977b; Winter et al. 1977; Winter and Troughton 1978; Winter 1981). *Calligonum* and many Chenopodiaceae are shrubby plants with highly reduced leaves and fleshy green internodes. These species possess palisade cortical chlorenchyma, continuous with leaf mesophyll (Metcalfe and Chalk 1950; Fahn 1963).

Structural studies have been published for the functionally aphyllous C_4 woody perennial *Hammada scoparia* with articulated, fleshy stems (Fahn 1963; Fahn and Dembo 1964; Kappen et al. 1975; Lyshede 1977b). The stem of this species has biseriate epidermis with thick-walled cells, covered by a relatively thin cuticle, and transversely oriented stomata (380 mm^{-2}) deeply sunken to the level of the inner layer. Thin-walled, uniseriate hypodermis encloses chlorenchyma consisting of two layers, outer palisade cells and shorter inner ones that function as leaf bundle sheath cells where 4-carbon acid decarboxylation and photosynthetic carbon reduction (PCR) occur (Fig. 4.13A). Nonphotosynthetic inner cortex is composed of highly vacuolate parenchyma, described as water-storing tissue, vascularized by a cortical vascular system. When photosynthetic activity of stem cortex is halted by extreme water stress (Kappen et al. 1975),

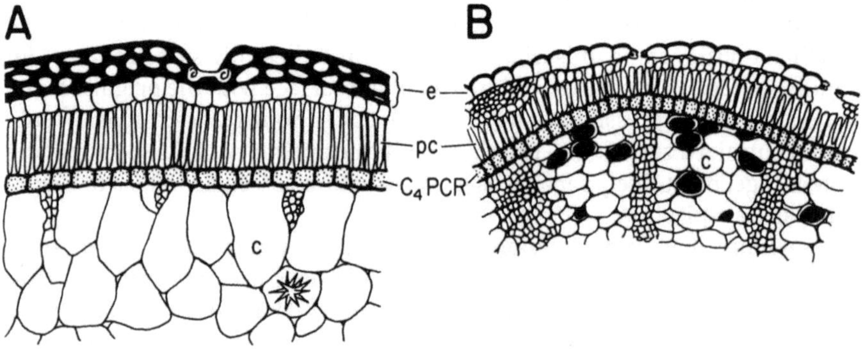

Fig. 4.13A,B. Transections of stems having C_4 photosynthesis. **A** *Hammada scoparia*. **B** *Calligonum comosum*. The layer where photosynthetic carbon reduction (PCR) cycle occurs is *stippled*; *c*, cortex; *e*, epidermis; *pc*, palisade chlorenchyma

the cortex is shed by formation of periderm (Fahn 1964), analogous to leaf dehiscence. Related desert species of *Hammada* and *Haloxylon* differ mainly in number and wall chemistry of epidermal layers, which for other species exhibit greater swelling when water is applied, probably attributable to abundance of hemicellulose and little pectin (Lyshede 1977b). This feature was hypothesized to be an adaptation to limit cuticular transpiration (Lyshede 1977a).

In *Anabasis articulata*, which has microphylls, stem anatomy is very similar to that of *Hammada scoparia* but has epidermal walls that rapidly imbibe water, lower stomatal densities (58 mm^{-2}), and much longer guard cells (Volkens 1887; Evenari 1938b; Fahn and Dembo 1964; Lyshede 1977a,b). Guard cells develop prominent wall thickenings during summer months, probably causing these to remain permanently closed (Gedalovich and Fahn 1983).

In *Calligonum comosum*, stem epidermis is uniseriate, has a relatively thin cuticle, and exhibits swelling of hemicellulose-rich cell walls. Thin-walled hypodermis has little contact with epidermal and palisade cells (Fig. 4.13B). Clusters of primary phloem fibers are well developed for leaf-trace bundles (Volkens 1887; Lyshede 1977b).

The annual species of *Salsola* also have C_4 chlorenchyma extending from leaves through stem internodes (Fahn and Broido 1963; Carolin et al. 1975, 1982). Stem angles are associated with diverging leaf-trace bundles and fortified with collenchyma; hence, cortical chlorenchyma is discontinuous and occupies less than half of the stem surface. Stems of *Salsola* and *Anabasis setifera* have one-layered epidermis with relatively thin cell walls, and do not develop sunken stomata (Lyshede 1977a). In fleshy-leaved species, the PCR sheath surrounds central mesophyll, which consists of water-storage parenchyma.

The C_4 desert chenopods differ markedly from the articulated C_3 species in *Salicornia, Arthrocnemum*, and *Allenrolfea*, which have a uniseriate epidermis but no hypodermis, multi-layered palisade chlorenchyma lacking bundle sheath-type cells, and special tracheids that radiate into palisade parenchyma from vein endings (Shields 1951a; Fahn and Arzee 1959; James and Kyhos 1961; Toelken 1967; Lyshede 1977a). Where the stem is constricted at each node, chlorenchyma is discontinuous.

4.3.3 Herbaceous Species

4.3.3.1 Solid Stems and Grass Culms

Little research has been conducted on the photosynthetic capacity of those herbaceous stems having significant stomatal conductance and formation of abundant chlorenchyma in outermost cortex.

Among herbaceous dicotyledons are *Linaria haelava* of the Negev, which has well-developed palisade chlorenchyma in the flowering stalk (Evenari et al. 1971) at a time when basal leaves have senesced and fruits are ripening. Similar examples occur in other Brassicaceae, e.g., *Brassica tournefortii* and species of *Sisymbrium* of North American deserts. Scape (stem) of numerous desert species of *Eriogonum* in western North America, such as *E. brachypodum* (Fig. 4.14) and *E. trichopes* (Went 1971) have unshaded palisade chlorenchyma and relatively high stomatal densities, the prerequisite anatomical features.

The Old World bunchgrass *Panicum turgidum* sheds its leaf blades but is described as maintaining green culms, but these instead may be leaf photosynthetic leaf sheaths, as they are in the North American C_4 *Pleuraphis rigida* (Chap. 3). Grass internodes typically have no or very few stomata and early sclerification of stem ground tissues.

4.3.3.2 Inflated Stems

Several herbaceous species have inflated green stems, whereby a greatly enlarged central cavity forms schizogenously from the pith, thereby producing a cylinder with increased photosynthetic surface area.

In North America the herbaceous perennial *Eriogonum inflatum*, known as desert trumpet (Fig. 4.15A), has a winter-spring basal leaf rosette and forms a green, spring-summer scapose inflorescence on which a fistula characteristically develops in the distal portion of the first scape internode, and sometimes progressively smaller fistulae form along that axis at higher internodes. Although insect larvae (Pyralidae) feed on the pith of these fistulae and thereby may increase size of the cavity (Stone and Mason 1979), swellings appear to be intrinsic, i.e., not caused by insect damage, and, moreover, are not uniformly expressed (Price 1982). Less pronounced swellings form on stems of *Eriogonum trichopes*.

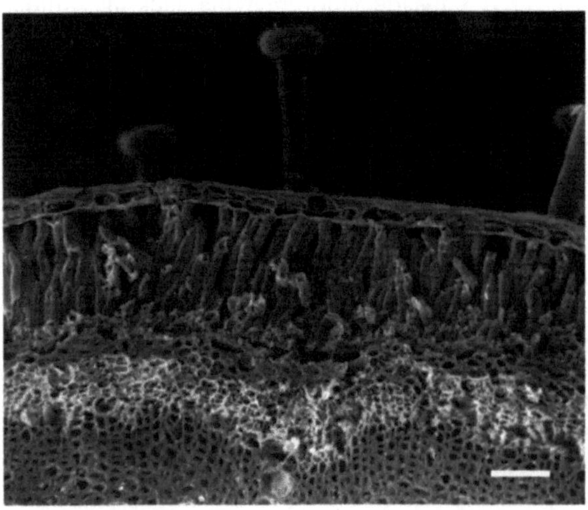

Fig. 4.14. *Eriogonum brachypodum*, a North American annual; transection of the photosynthetic stem of the inflorescence, showing well-developed palisade chlorenchyma in cortex. *Bar* 50 µm

In *E. inflatum*, scape photosynthesis constitutes an important adaptation for this species. In Death Valley, scape area per plant ranged from 80% on the largest plant to less than 50% on smaller plants, and stem photosynthesis yielded up to 66% of yearly carbon fixation (Smith and Osmond 1987). Rosette leaves had maximum CO_2 uptake of 20–23 μm m^{-2} s^{-1} and relatively high transpiration rates as compared with stems having CO_2 uptake of 12–17 μm m^{-2} $s^{-}1$ with more conservative instantaneous WUE than leaves (Osmond et al. 1987). During hot days, near-soil leaves were 4–8 °C above air temperature and therefore increased to lethal temperatures, and leaf rosettes declined when midday water potentials were less than −2.0 MPa; hence, leaves senesced before summer. A scape that by its distance from the soil and vertical orientation was cooler avoided lethal temperatures and survived into the dry season long after leaves senesced, even persisting when predawn water potential was less than −3.0 MPa (Smith and Osmond 1987).

A basal leaf of *E. inflatum* has scattered adaxial trichomes but relatively dense abaxial ones, which appear to be reflectant, presumably to minimize heating from the soil but probably reducing PPFD interception as well. These amphistomatic and isolateral leaves have thin cuticle. Scapes are glabrescent and have significantly lower stomatal densities, larger stomatal pores, thicker cuticle, and lower A^{mes}/A values due to thinner palisade layer. The scape has only half the chlorophyll content of a comparable leaf surface, and its light and CO_2 saturation are likewise significantly less (Osmond et al. 1987). Thus, a scape has a less efficient design in its photosynthetic apparatus than in basal leaves,

Fig. 4.15A,B. Inflated photosynthetic stems from western North America. A *Eriogonum inflatum*. B *Caulanthus inflatus*

excepting perhaps percent PPFD absorptance, which needs to be measured. CO_2 concentration in the scape cavity was measured as high as 3880 μbar, and this internal supply, undoubtedly resulting from tissue respiration, appears to be partially recycled and possibly could sustain stem photosynthesis during daytime water stress when stomata close (Osmond et al. 1987).

In the robust winter annual *Caulanthus inflatus* (Fig. 4.15B), the erect yellow stem has a large hollow center while also forming conspicuous cauline leaves. Leaves, stems, and developing fruits, all of which have stomata, have substantial rates of CO_2 uptake (Gibson and Sharifi unpublished data). Stem stomatal densities and A^{mes}/A values are extremely low relative to those of cauline leaves, and its chlorenchyma does not differentiate as palisade layers.

When young and soft, the hollow stems of both species may become lodged during moderate winds, but when hard and dried, they often persist for months. Mechanical studies are needed to determine how such club- or double-tapered cylinders may be designed to minimize damage from bending and torsion.

5 Succulent Photosynthetic Organs

A succulent photosynthetic organ typically maintains positive daily carbon balance by having its own water reserve and accompanying high organ water potential. The organ possesses anatomical and physiological adaptations that yield very high WUE, even when it is severed from the root system (Nobel 1985, 1988; Gibson and Nobel 1986). Succulence grades into "fleshiness," which describes leaves having high water capacitance but wilting rapidly when a shoot is severed (Gibson 1982); fleshy organs lack many mechanisms of drought resistance and in warm deserts tend to be drought-deciduous structures or occur on short-lived annuals. Therefore, succulence cannot be judged only by measuring the amount of stored water or thickness of the organ (von Willert et al. 1992).

Although many authors have described desert succulents as drought tolerant, in the terminology of Levitt (1980) a succulent leaf or stem may be described both as drought-avoiding, i.e., using adaptations to maintain higher tissue water potential than in the soil, and drought-tolerant, i.e., remaining alive when cell water content is low. Some succulents can tolerate water content as low as 20% and still show cyclic rehydration and retention of carbon assimilation (Gibson and Nobel 1986; Nobel 1994a). Even water-stressed succulents tend to maintain water potentials in a persistent photosynthetic organ higher than −2.0 MPa.

In warm deserts, many succulents are heat-tolerant, on hot days storing heat that would be lethal to typical plants (Nobel et al. 1986; Eller and Grobbelaar 1986), and not cold-resistant, but some species utilize morphological or biochemical adaptations to maintain tissue temperatures above freezing (Nobel 1980a,b,c, 1984; Loik and Nobel 1991; Goldstein and Nobel 1991, 1994). Typical succulents are also salt-escapers, whereas many fleshy desert plants are xerohalophytes and thereby salt-resisters, using either avoidance or tolerance mechanisms (Waisel 1972; Reimold and Queen 1974; Sen and Rajpurohit 1982; Staples and Toenniessen 1984). Numerous Aizoaceae accumulate NaCl and are exceptionally salt-tolerant but are not halophytes (von Willert et al. 1992).

5.1 General Properties of Succulent Photosynthetic Organs

Warm desert plants with succulent photosynthetic organs belong to approximately 20 different angiosperm families (Table 5.1), and for these it is difficult

Table 5.1. Representative succulents in warm deserts of the world

Families with CAM
Agavaceae: CAM species *Agave cerulata, A. deserti, A. lechuguilla, A. sobria, Yucca torreyi*; C_3 species *Y. brevifolia, Y. schidigera*
Aizoaceae: *Cephalophyllum* subg. *Homophyllum, Conophytum* spp., *Dinteranthus wilmotianus, Drosanthemum floribundum, Fenestraria aurantiaca, Lithops* spp., *Psilocaulon marlothii*; facultative CAM *Mesembryanthemum crystallinum, M. forskallii*
Asclepiadaceae (Stapelieae): *Caralluma negevensis C. fenestrata, C. deflersiana, C. commutata, Duvalia polita, Echidnopsis bavazzanii, E. bihendulensis, E. squamulata, E. virchowii, Hoodia bainii, H. currori, Huernia marnieniana, H. oculata, Orbea rangeana, Piaranthus ruschii, Psuedolithos* spp., *Rhytidocaulon macrolobum, Stapelia flavopurpurea, S. remota, S. schinzii, Stultitia araysiana, Tavaresia angolensis, T. barklyi, Trichocaulon clavatum, T. meloforme, T. triebneri, Tridentea ruschiana*
Asphodelaceae: *Aloe asperifolia, A. dhufarensis, A. dinteri, A. hereroensis, A. inermis, A. karasbergensis, A. megalacantha, A. namibensis, A. rubroviolacea, A. scorbinifolia, Haworthia venusa*
Asteraceae: *Kleinia deflersii, K. neriifolia, K. subulifolia*
Bromeliaceae: *Tillandsia* spp., *Hechtia* spp.
Cactaceae: *Carnegiea gigantea, Copiapoa cinerea, C. haseltoniana, Echinocactus polycephalus, Echinocereus fasciculatus, Echinopsis leucantha, Eulychnia iquiquensis, Ferocactus acanthodes, F. covillei, F. fordii, Lophocereus schottii, Lophophora williamsii, Mammillaria poselgeri, Neoraimondia macrostibas, Opuntia basilaris, O. bigelovii, O. fulgida, O. leptocaulis, O. pampeana, O. ramosissima, Pachycereus pringlei, Peniocereus johnstonii, Pterocactus tuberosus, Stenocereus alamosensis, S. eruca, S. gummosus, S. thurberi, Trichocereus candicans, T. coquimbanus*
Crassulaceae: *Crassula orbiculata, Dudleya saxosa, Tylecodon paniculatus, T. pearsonii, T. wallichii*
Euphorbiaceae: *Euphorbia abdelkuri, E. abyssinica, E. antisyphilitica, E. atrox, E. awasmontana, E. ballyi, E. balsamifera var. adenensis, E. cameronii, E. caducifolia, E. chersina, E. damarana, E. dregeana, E. fascicaulis, E. fusca, E. gillettii, E. guerichiana, E. horwoodii, E. hypogaea, E. juttae, E. lignosa, E. multiclava, E. namibensis, E. reptans, E. schimperi, E. socotrana, E. spinea, E. turbiniformis, E. verrucolosa, E. virosa, E. yemen, Pedilanthus macrocarpus*
Geraniaceae: *Pelargonium crithmifolium, P. desertorum*
Portulacaceae: *Anacampseros alstonii, A. comptonii, A. crinita, A. karasmontana, A. namaquensis; Ceraria fruticulosa, Portulacaria armiana, P. pygmaea; Portulaca foliosa* (C_4)
Pachycaulous and caudiciform succulents with nonsucculent leaves (C_3, no CAM)
Anacardiaceae: *Pachycormus discolor*
Apocynaceae: *Adenium boehmianum, A. obesum, A. somaliense, Pachypodium namaquanum*
Burseraceae: *Bursera hindsiana, B. microphylla, Commiphora dinteri, C. multijuga, C. opobalsamum, C. virgata*
Cucurbitaceae: *Dendrosicyos socotrana, Ibervillea sonorae*
Fouguieriaceae: *Fouguieria burrager, F. columnaris, F. shrevei, F. splendens*
Geraniaceae: *Sarcocaulon inerme, S. flavescens, S. mossamedense, S. multifidum, S. patersonii*
Moraceae: *Dorstenia gigas*
Moringaceae: *Moringa ovalifolia*
Passifloraceae: *Adenia ballyi, A. pechuelii*
Pedaliaceae: *Sesamothamnus guerichii*
Vitaceae: *Cyphostemma currorii* (large, thick leaves)

to generalize without finding occasional exceptions. However, the following are common features of these photosynthetic organs.

1. Water content is 90% or greater in a fully hydrated organ.
2. The succulent organ has a relatively high volume-to-surface ratio and possesses thick parenchymatous tissues for water storage.
3. Crassulacean acid metabolism (CAM) is the principal mode of carbon assimilation in the majority of species.
4. Daytime water vapor conductance is extremely low.
5. Stomata are relatively large and occur on most exposed green surfaces but at fairly low densities.
6. Each chlorenchyma cell tends to have a large volume, includes a central vacuole, which occupies greater than 90% of the cell, and has a relatively thin cell wall; turgor of chlorenchyma is preferentially maintained from water of adjacent achlorophyllous ground tissue.
7. Succulent organs that maintain high turgor pressure mechanically behave as hydrostats.

5.1.1 Water Content

Water may constitute 90% or more of a fully hydrated photosynthetic organs, e.g., succulent stems of many Cactaceae, *Euphorbia*, and Asclepiadaceae, or succulent leaves of many Aizoaceae, *Aloe*, and Crassulaceae. Stems with substantial woody cylinders, as in the columnar cactus *Carnegiea gigantea* and platyopuntias (*Opuntia* spp.), and leaves of *Agave* and *Yucca*, which have abundant bundle fibers, have lower saturated water content due to a greater proportion of dry biomass, but also possess very large tissue volumes for water storage. In fact, a fibrous network is needed to support large, heavy photosynthetic organs in *Agave* and *Yucca* (Nobel 1988, 1994a) and cladodes of platyopuntias (Gibson and Nobel 1986). After being water-saturated, a succulent organ may experience daily transpiration for 1 to 2 months and still have higher water content than many fully hydrated nonsucculent leaves.

5.1.2 High Volume-to-Surface Ratio

Some measure of organ water content traditionally has been used to define succulence. Delf (1912) defined *degree of succulence* as water content at saturation (g) per surface area of the organ (dm^{-2}). If water content is measured instead as volume (mm^3), then the relationship becomes volume of water/surface area ($mm^3\ mm^{-2}$, V/A), which is one way to view the importance of succulence as volume of water stored behind each sector of transpiring surface (Gibson and Nobel 1986; Nobel 1988). Nobel used V/A to estimate duration of stomatal opening, which in cacti and agaves increases as *V/A* increases. Lower *V/A* ratios in seedlings of succulents versus older plants, hence lower relative water storage, account for death due to desiccation because seedlings cannot tolerate long droughts (Jordan and Nobel 1981; Gibson and Nobel 1986). For adult plants of *Agave*, *V/A* ratios of leaves increase with increasing aridity (Burgess 1985).

During drought, while shoot water content gradually decreases, organ shape changes and cross-sectional area decreases as cell turgor decreases. For example, in succulent leaves from the Karoo, cross-sectional area decreased 27–60% from turgid to dry state, produced by similar losses in water volume from mesophyll (von Willert et al. 1992).

Volume-to-surface ratio of a sphere (*r*/3, where *r* is the radius), is greater than that of a cylinder (*r*/2), neglecting the ends, so that a sphere has more water-storing volume per unit surface area (Gibson and Nobel 1986). Conversely, projecting tubercles (e.g., in Cactaceae and Stapelieae), ribs, or grooves on a sphere or cylinder decrease that value, thereby increasing surface area and potentially increasing photosynthetic area and growth rate. Thus, area required for photosynthesis and volume for water storage can be viewed as opposing strategies.

Ribbing on stem succulents permits a stem to expand and contract with changes in hydration but without causing physical damage to cells (Spalding 1905; MacDougal and Spalding 1910; Gibson and Nobel 1986). Especially in Cactaceae, flexibility of organ covering is facilitated by presence of collenchymatous hypodermis.

Computer simulations tested effects of rib number on PPFD interception and CO_2 uptake by ribbed columnar cacti with relatively deep versus shallow ribs (Geller and Nobel 1984). For plants with increasing number of shallow ribs, such as barrel cacti (*Ferocactus*) or *Carnegiea gigantea*, total daily PPFD and CO_2 uptake per surface area decreased but stem productivity was greater with more ribs because total stem area increased (Fig. 5.1). For plants with increasing number of deep ribs, such as *Lophocereus schottii*, plant productivity increased to 13 ribs and then decreased. Assuming similar responses in succulent euphorbias, a curvilinear increase in plant productivity would be expected in changing from cylindrical to few-ribbed stems.

For cacti, CO_2 uptake on clear days should be maximal when stem perimeter divided by diameter (perimeter ratio) equals about 6, roughly double that of a cylinder, regardless of rib number but should be close to 4 whenever shading is common (Geller and Nobel 1984). Under field conditions, where some self-shading occurs in typical Sonoran Desert habitats, three species of columnar cacti had perimeter ratios of 4.1–4.4.

Ribs tend to limit the percent surface area that is PPFD-saturated but are benefited by trapping enough reflected PPFD on shaded surfaces to minimize stem area below PPFD compensation. Ribbing has very small daytime thermal effects (Lewis and Nobel 1977). Tubercles on ribbed or unribbed succulents increase photosynthetic surface, and cladodes result in low *V/A* ratios and tend to be oriented to maximize PPFD interception (Nobel 1982a,b; Gibson and Nobel 1986). In the open coastal plain of the Atacama, *Copiapoa haseltoniana*,

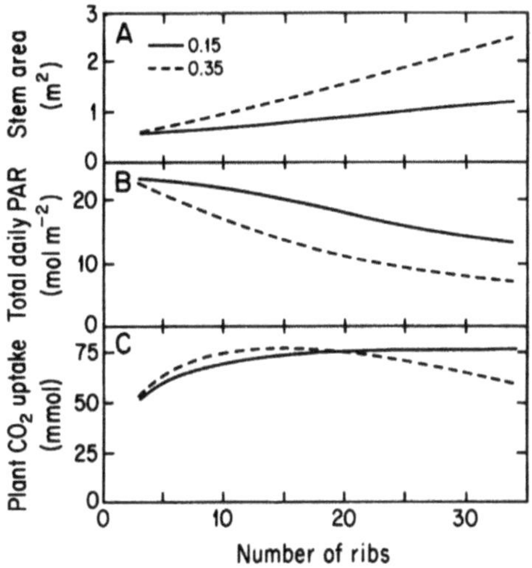

Fig. 5.1A–C. For cacti have deep ribs (——) or shallow ribs (-----), influence of rib number on stem surface (**A**), average total daily PAR (PPFD) per unit stem area (**B**), nocturnal CO_2 uptake for the whole plant (**C**). (Gibson and Nobel 1986)

a low ribbed cactus, is oriented to minimize absorption of solar radiation but to raise apical temperature for increasing vegetative growth and flowering (Mooney et al. 1977c; Ehleringer et al. 1980).

Whereas succulents are characteristically described as having low surface-to-volume ratios (Smith and Nobel 1986) or low surface expansion ($mm^2 g^{-1}$ fresh weight; von Willert et al. 1992), their photosynthetic organs have epidermis composed of groundmass cells that often are convex or papillose. Cell surfaces and waxy microstructures can produce substantial enlargement of surface area, and thus may have an effect on turbulence within the boundary layer and play a roll in heat transfer (Porembski et al. 1991). Nonetheless, the typical effects of an irregular organ surface and thick wax most likely are to reflect solar radiation, especially heat-producing IR, and to absorb UV (Mulroy 1979). Whether papillose micromorphological relief has an adaptive function in heat transfer, as in subterranean stems of *Pterocactus tuberosus* and *Peniocereus johnstonii* (Cactaceae; Gibson 1978a; Gibson and Nobel 1986) and buried leaves of certain *Cheiridopsis* and *Lithops* (von Willert et al. 1992), remains to be experimentally studied.

5.1.3 Crassulacean Acid Metabolism

Carbon assimilation in most succulents uses Crassulacean acid metabolism (CAM), whereby carbon uptake occurs mainly at night, when temperatures are low, and stomata are closed during daylight, when temperatures are high. Stomatal opening at night results in tremendous water conservation because Δc_{wv} [Eq. (3.5)] is smaller when temperature is lower, thus water vapor concentration of the air is low, and transpiration may be decreased by 100% (Gibson and Nobel 1986; Nobel 1988).

CAM has been studied intensively for 20 years, and cellular and biochemical events have been discussed in detailed reviews (Kluge and Ting 1978; Osmond 1978; Kluge 1979; Osmond et al. 1982; Ting and Gibbs 1982; Gibson and Nobel 1986; Nobel 1988, 1991b). When stomata open at night, CO_2 diffuses into chlorenchyma where it combines with phosphoenolpyruvate (PEP) in the cytosol to form an organic acid, typically malate. Malate and related organic acids are then loaded into the vacuole (Fig. 5.2). As malate accumulates in vacuoles, tissue pH decreases during the night until an equilibrium is reached. At sunrise, stomata begin to close, although many CAM plants exhibit a short burst of C_3 photosynthesis just after daybreak. During daylight, when stomata are closed, malate is returned to cytosol and there decarboxylated; CO_2 then diffuses into chloroplasts, where it is combined with ribulose 1,5-biphosphate for the dark reaction of Calvin-Benson photosynthesis.

CAM has evolved convergently in Cactaceae, Agavaceae, Aizoaceae, Asphodelaceae, Crassulaceae, Asclepiadaceae, and Euphorbiaceae, families in which C_3 photosynthesis is the plesiomorphic state.

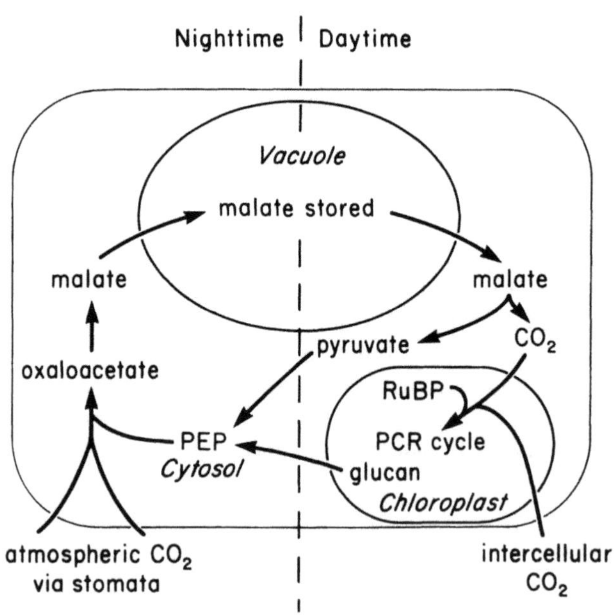

Fig. 5.2. Basic biochemical events within a cell having Crassulacean acid metabolism. At nighttime, atmospheric CO_2 diffuses through stomata and in liquid phase enters cytosol, where CO_2 reacts with PEP (phosphoenolpyruvate) to form oxaloacetate, rapidly converted into malate or other organic acids, which are actively transported into the central vacuole. During daylight, malate is removed from the vacuole and decarboxylated, releasing CO_2 to be taken up by chloroplasts for photosynthetic carbon reduction (PCR cycle) when combined with RuBP (ribulose 1,5-bisphosphate). From the decarboxylation, pyruvate can be used to make more PEP. Respiratory CO_2 and CO_2 released from other cells are also sources for this cell

5.1.4 Epidermal Water Vapor Conductances

Photosynthetic succulent organs generally retain an intact epidermis, including fully functional stomata. This tissue is highly effective in limiting transpiration during daytime, when stomata are in a closed mode for CAM, because cuticle of these species is either of high quality or very thick or both. Daytime porometric readings for CAM and C_3 succulents often record extremely low cuticular conductances, 0.01–0.1 mm s^{-1}, values less than for nonsucculent plants (Nobel 1991a).

At night, when stomata are wide open, characteristic stomatal conductances of CAM succulents are also much lower than those of nonsucculent plants. For example, nocturnal water vapor conductances were 0.6–3.0 mm s^{-1} (25–120 mmol H_2O m^{-2} s^{-1}) for *Agave deserti, Ferocactus acanthodes*, and *Carnegiea gigantea* (Nobel 1988), whereas daytime water vapor conductances for mesophytes typically were 4–20 mm s^{-1} (Nobel 1991a). Virtual absence of transpiration during daytime eliminates evaporation as a cooling mechanism for succulent organs; thus heat is stored and builds up during midday heat and then is gradually lost at night (Mozingo and Comanor 1975). On a hot day, surface layers, including chlorenchyma, may greatly exceed ambient as heat is stored in the stem, achieving higher temperatures than core tissues of thick organs (pith, inner cortex, or central mesophyll), which, conversely, take longer to cool.

Although the majority of warm desert succulents have a glabrous epidermis, lacking trichomes, presumably to maximize absorptance of PPFD by chlorenchyma, many species of *Haworthia*, including the deserticolous *H. venosa* subsp. *tessellata*, have whitish warty spinules on the leaf, actually small domes with thick-walled cells formed by proliferation of protoderm (Gibson 1982).

The surface cover (skin) of a cactus stem consists of epidermis impregnated with wax and several layers of living collenchymatous hypodermis, which has substomatal canals through which gases diffuse between the air and chlorenchyma (Fig. 5.3; Gibson and Horak 1978). In the giant cactus *Carnegiea gigantea*, a spectral analysis of skin 1 mm thick with abundant wax revealed that this surface covering was opaque to UV radiation and decreased transmittance of PPFD by 36% of incident light reaching chlorenchyma (Darling 1989). Presence of PPFD-absorbing skin contributes to lowered net primary productivity, given that cacti are strongly PPFD-limited in an environment where full sunlight prevails (Nobel 1980c; Gibson and Nobel 1986). Absorptance of UV is probably accomplished by presence of flavonoids in these surface layers (Robberecht and Caldwell 1978; Gibson et al. 1986; Darling 1989), thereby protecting chlorenchyma from high-energy damage.

5.1.5 Stomatal Patterns

Succulents have stomata on all green surfaces (opposite chlorenchyma), but these occur at densities lower than on nonsucculent xerophytes and mesophytes,

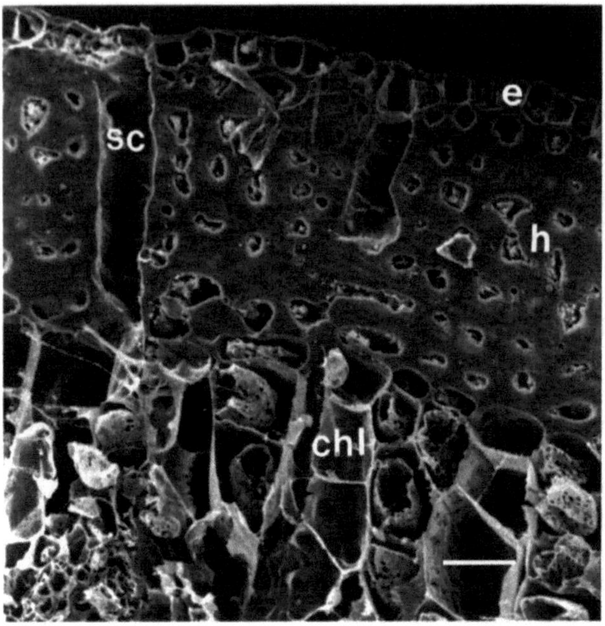

Fig. 5.3. Transection of outermost stem of *Stenocereus gummosus*, showing tough skin composed of epidermis (*e*) and thick-walled hypodermis (*h*) interrupted by substomatal canals (*sc*), leading to chlorenchyma (chl) of the outer cortex. *Bar* 50 µm (Gibson and Nobel 1986)

for warm desert leaf and stem succulents rarely exceed 60 stomata mm^{-2}. Guard cells on succulents are relatively large, thus creating large stomatal pores (Zemke 1939; Sundberg 1985, 1986), but overall a succulent photosynthetic organ has low stomatal area per unit surface area (A^{st}/A). Low A^{st}/A, sometimes less than one-tenth of that in other plant forms, may be the most important factor contributing to low stomatal conductances in succulents, which likewise are an order of magnitude less than in nonsucculent leaves.

Succulent leaves and stems have great surface area, and to accommodate expansion, some succulent dicotyledons utilize protracted division of protodermal cells to produce extra cells for the mature epidermis. In Cactaceae from deserts and other dry habitats, groundmass epidermis of stem ribs and tubercles commonly displays polygonal arrays of several cells (to 20), each array derived from a protodermal mother cell (Fig. 5.4). Polygonal arrays also occur on stem ribs of many species of *Euphorbia*. The presence of neither large groundmass epidermal cells nor polygonal arrays is an exclusive feature of desert species and appears to have evolved in ancestral semiarid habitats.

Several lineages with succulent photosynthetic organs, especially in Crassulaceae, initiate stomatal apparati throughout organ expansion, so that new stomata arise among old ones, and succulent dicotyledons often form mesogenous parallelocytic or helicocytic stomata, whereby a single meristemoid produces an extended cluster of subsidiary cells (Payne 1970).

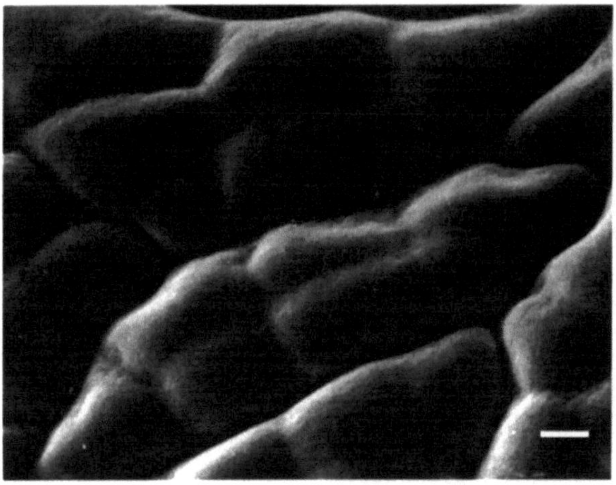

Fig. 5.4. *Stenocereus gummosus*; SEM view of stem surface composed of polygonal arrays of groundmass cells, each array derived from one protodermal initial. *Bar* 25 μm

Succulent stems may remain green and thereby photosynthetic for many years because periderm formation is greatly delayed, thus preserving epidermis with fully functional stomata. Periderm arises superficially and commonly first appears as patches in response to sunscalding, freezing, or mechanical damage (Gibson and Nobel 1986).

5.1.6 Chlorenchyma

Chlorenchyma, which in any fully exposed, thick organ has centric (continuous) arrangement, consists mostly of wide cylindrical to isodiametric or prolate spheroidal cells (Fig. 5.5) rather than neatly layered, narrow palisade parenchyma of nonsucculent photosynthetic organs. Chlorenchyma may be as thick as 3 mm, but rarely exceeds 1 mm (Gibson 1982). Each cell contains a large central vacuole; this is a key component for CAM species, in which quantity of acid stored at night is linked to total sugar production during daylight (Nobel 1988). Presence of such conspicuous vacuoles and large intercellular air spaces results in comparatively few chloroplasts per surface area, sometimes expressed as high mesophyll succulence (g fresh tissue mg^{-1} chlorophyll; Kluge and Ting 1978).

Two types of leaf succulence, all-cell succulent and partially succulent leaves, have been recognized (von Willert et al. 1992). All-cell succulent leaf describes the case where mesophyll is homogeneous, so that each parenchyma cell is photosynthetic and also stores water, whereas the partially succulent type has a distinct achlorophyllous water-storing tissue, in which volume per cell may be as much as 10^3 greater than a chlorenchyma cell and the vacuole occupies more than 99% of each cell. Very few species, basically only those with thin leaves, would

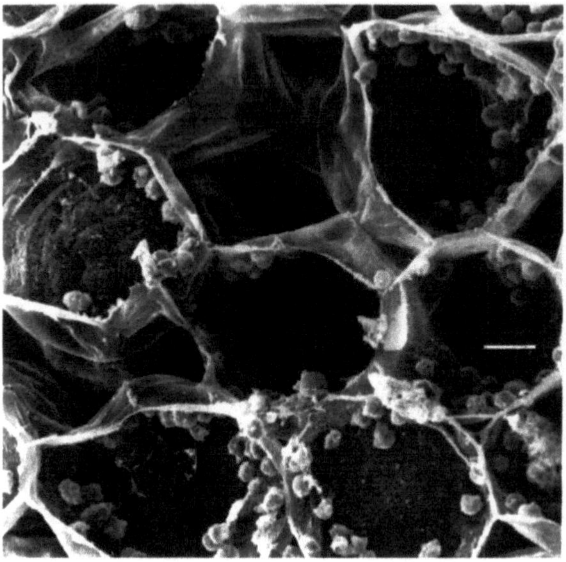

Fig. 5.5. Photosynthetic stem cortex of *Stenocereus gummosus*, with large, thin-walled, isodiametric cells, each of which has peripheral chloroplasts and a very large central vacuole. *Bar* 10 μm (Gibson and Nobel 1986)

fit the definition for all-cell succulence, and these are not those well adapted to very long droughts. The majority of species would be thick leaves of the partially succulent type, including those in which the transition from chlorenchyma to achlorophyllous tissue is gradual and also leaves with a sharp boundary between peripheral leaf chlorenchyma and central ground tissue, e.g., in *Aloe*.

Achlorophyllous mesophyll or inner stem cortex serves as a specialized water-storing tissue in that during severe dehydration these cells preferentially lose water, while osmotic mechanisms are present that maintain higher turgor potential in chlorenchyma to preserve its function (Barcikowski and Nobel 1984; Schmidt and Kaiser 1987; Schulte and Nobel 1989; Schulte et al. 1989b; Goldstein et al. 1991a; Tissue et al. 1991). Because succulents open stomata during periods when soils are dry, hence no water can be delivered directly via xylem from roots to photosynthetic tissues, large volumes of stored water in achlorophyllous parenchyma provide much of the water transpired (Smith et al. 1987; Schulte et al. 1989b). Then, when stomata are closed, there is a daily redistribution of internal water between storage cells and chlorenchyma, driven by changes in osmotic pressures. Especially chorenchyma cells with high nocturnal osmotic pressures, resulting from CAM manufacture and storage of organic acids, maintain turgor by receiving water from adjacent storage tissues. During long drought, turgor pressure in chlorenchyma remains high, while being greatly decreased in storage parenchyma. For an all-cell succulent leaf to maintain leaf turgor, stems would have to be closely involved in diel water movement.

Complex mucopolysaccharides comprise mucilage, which is produced within large idioblasts of ground tissue in many desert succulents, but often is then excreted into the apoplast (Trachtenberg and Fahn 1981; Gibson and Nobel 1986). In mucilaginous stems of three desert species of cacti, mucilage content averaged 23% dry weight in chlorenchyma and 28% in adjacent water-storing parenchyma (Nobel et al. 1992a). Whereas rehydration of a mucilaginous stem was slower than in *Ferocactus acanthodes*, which lacks mucilage, mucilage appeared to function as a passive apoplastic capacitor, enabling the plant to store additional water while helping to buffer parenchyma from short-term water deficits.

In general, mesophyll and watery cortex of succulent organs lacks secondary cell walls, permitting cells to collapse during dehydration and thereby function like hinges (Barcikowski and Nobel 1984; Schulte and Nobel 1989; von Willert et al. 1992). Preferential hydration of chlorenchyma during organ shrinkage preserves the integrity of that tissue. Special adaptations are present in some succulent organs to minimize physical damage to other cells caused by profound dehydration. In cacti, helical secondary thickenings in elements of secondary xylem permit these stems to contract axially (Gibson 1973). Species of *Sansevieria* have special secondary thickenings in water-storage cells to minimize their physical damage (Koller and Rost 1988a,b).

5.1.7 Succulent Hydrostats

In a biomechanical sense (Niklas 1992), many succulent organs behave as a hydrostat, having massive thin-walled parenchyma tissue capable of generating high internal hydrostatic pressure uniformly applied on a shear-resistant outer surface. Especially if the water-storing parenchyma has relatively low volume of intercellular air spaces and is turgid, bulk modulus is high, and the core can function as an incompressible structure. The surface layer, which must consist of strong tensile materials, can yield a firm structure with little cost invested.

Although a shear-resistant cover is a key component in maintaining the shape and orientation of the organ, the same tough epidermis or complex surface tissue simultaneously also would lower cuticular water vapor conductance, filter or reflect some harmful solar radiation from reaching chlorenchyma, and deter some invertebrate herbivory.

5.2 Taxon-Specific Adaptations

5.2.1 Aloe

Aloe, a CAM genus with nearly 300 species in semiarid eastern and southern Africa, is also well represented in Old World warm desert habitats. Aloe leaves, which characteristically have very thick outer periclinal epidermal cell walls but nearly always lack fibers, are prime examples of hydrostats (Beaumont et al. 1985), wherein leaves are held vertical to horizontal using turgor within a tightly sealed structure. Turgor is maintained within a nearly transparent, central water-storing parenchyma, composed of huge cells lacking plastids and starch and having very small intercellular air spaces (Fig. 5.6). By using this mechanism, young vertical leaves of *A. dichotoma* received only 62% of total solar radiation intercepted by old horizontal leaves (Flach 1986). The same mechanism probably controls leaf orientation in most other Alooideae, which also lack leaf fibers (Beaumont et al. 1985).

Cell wall and surface wax on leaves of warm desert species combined has a thickness of at least 20 μm (Volkens 1887; Smith and Van Wyk 1992). Groundmass epidermis is papillose in many species, e.g., *A. dhalensis, A. dichotoma, A. ramosissima, A. namibensis* (Glen and Hardy 1986), *A. asperifolia*, and *A. hereroensis* (Fig. 5.7A), *A. karasbergensis* has numerous micropapillae per cell (Fig. 5.7B), and *A. scobinifolia* has a verrucose surface resulting from outward elongation of the cells (Fig. 5.7C). Surface rugosities and thick waxes account for the grayish appearance of leaves and the formation of deeply sunken stomata, concealed by overarching lobes of four subsidiary cells (Cutler 1972; Brandham and Cutler 1978), recessed the thickness of the wall to 75 μm, thus creating a rectangular outer aperture (Fig. 5.7D) and a suprastomatal cavity that is long enough likely to cause some slowing of water vapor and CO_2 diffusion at high stomatal conductances (Gibson 1983). In *A. scobinifolia* (Fig. 5.7C), sunken stomata occur along grooves between groundmass cell clusters. Inner periclinal cell walls of epidermal cells are thickened, as much as 15 μm in *A. ramosissima,* adding to the mechanical toughness of the epidermis to contain heavy water tissues and probably further restrict epidermal water vapor conductance.

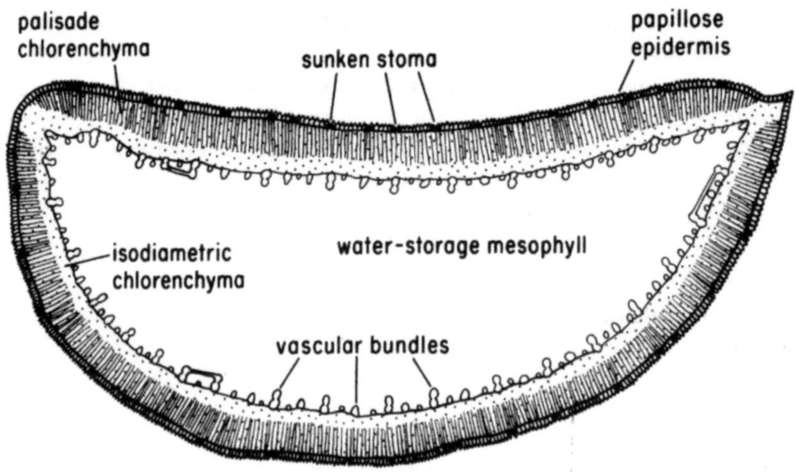

Fig. 5.6. Leaf structure of *Aloe namibensis*, an example of a hydrostat

Chlorenchyma is centrically arranged, sharply demarcated by a ring of vascular bundles, up to 1.4 mm thick (*A. namibensis*), and composed of numerous layers of short cylindrical cells with concentrated chlorophyll inwardly grading to isodiametric cells with less chlorophyll. All individual cells may round with age. Typically, chlorenchyma cell walls are somewhat thickened, especially as they age, sometimes exceeding 2 μm, signifying that liquid-phase CO_2 diffusion may be slowed as a cost of providing some rigidity for the outer leaf.

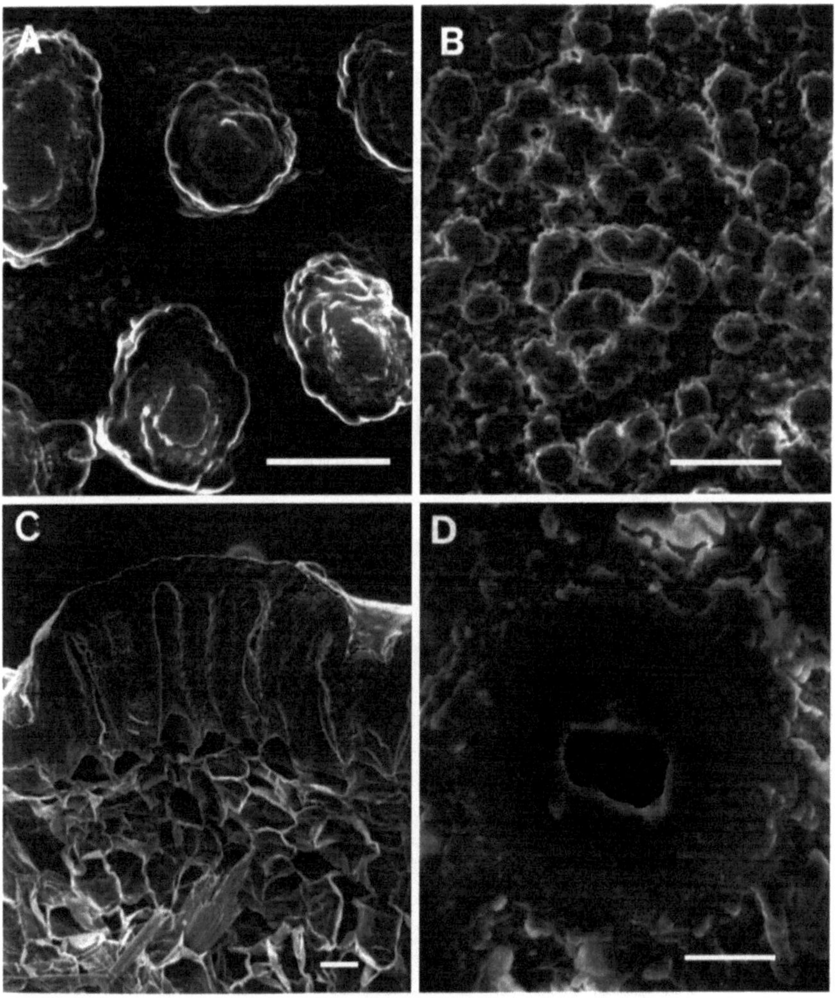

Fig. 5.7A–D. Leaves of *Aloe*. **A** *A. hereroensis*, surface with papillae. **B** *A. karasbergensis*, surface with micropapillae. **C** *A. scobinifolia*; transection showing elongation of epidermal cells to produce rugose surface. **D** *A. peirsonii*; rectangular aperture overarching sunken stoma. *Bar* A–C 25 μm; **D** 10 μm

5.2.2 Agave and Yucca

In sharp contrast to *Aloe*, succulent leaves of North American *Agave*, a CAM genus, and *Yucca*, in lowland deserts mainly a C_3 genus, do not behave as hydrostats in that they possess an elaborate system of scattered fibrous vascular bundles, used for this reason as commercial fiber (Schery 1972; Smith et al. 1987; Nobel 1994a). Agavoid leaves apparently utilize great tensile properties of fiber strands. Agaves characteristically form basal rosettes in which the fibrous leaves are oriented in a helically alternate phyllotaxis that maximizes daily and seasonal PPFD interception while minimizing leaf overtemperature to less than 7 °C (Woodhouse et al. 1983).

Warm desert Agavaceae, like those in other habitats (Blunden et al. 1973), have thick epidermal outer cell walls with much wax (Fig. 5.8A), a minimum of 20 μm combined thickness (Bray 1903). Groundmass epidermis may be relatively flat, e.g., in *A. shawii, A. striata, Y. valida*, papillose as in *A. deserti, A. cerulata*, and *Y. schidigera*, or with a highly irregular surface resulting from outward elongation of cells, as in *A. sobria* (Fig. 5.8B). Grayish leaves typically have epicuticular wax coating prominent papillae. Epidermal cells are radially elongate and have domed secondary walls with flanges to at least halfway down the anticlinal walls, e.g., in *A. shawii* and *A. lechuguilla*, but may extend almost to the base, e.g., in *A. striata* and *Y. schidigera*. Inner periclinal walls are generally thin but show some thickening in a few species of both genera.

Guard cells remain at the same level as the unthickened base of groundmass cells, hence they become sunken by the outward elongation of surround-

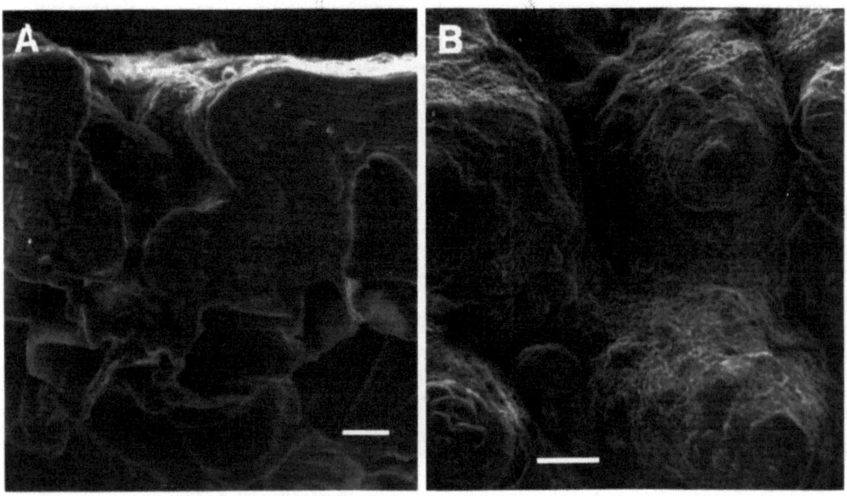

Fig. 5.8A,B. Leaves of desert *Agave*. **A** *A. lechuguilla*; transection of epidermal cells with very thick, cutinized cell walls, producing a deeply sunken stoma. **B** *A. sobria*; microrelief consists of conic groups of cells, between which are located sunken stomata. *Bar* 5 μm; **B** 50 μm

ing epidermis with secondary wall formation and cuticle development. Viewed from the surface, a suprastomatal cavity becomes partially closed by two pairs of cells surrounding the guard cells that overarch and thereby protrude into the cavity (Gentry and Sauck 1978). Although this produces a labyrinth effect, overall width of the suprastomatal pathway probably is greater than stomatal aperture, and any reduction in stomatal conductance is most significantly affected therefore by length of the cavity [Eq. (3.9)].

Between epidermis and outermost vascular bundles occurs chlorenchyma rich in chloroplasts. These cells are short cylindrical cells in most species but isodiametric in leaves where bundles are close to the epidermis, e.g., *Agave striata*, *Yucca schidigera*, and *Y. valida*. Nonchlorophyllous mesophyll may have large intercellular air spaces, as in *Y. schidigera* and *Y. brevifolia*, but these cells may also develop relatively thick cell walls.

A leaf of *A. deserti* has eight to ten layers of chlorenchyma outside vascular bundles, less dense chlorophyll in inner chlorenchyma among vascular bundles, and central water-storage parenchyma, all of which vary in cellular features along the leaf axis (Smith and Nobel 1986; Smith et al. 1987). Parenchyma volume and total cell area increased toward the base, where eight to nine cells separated bundles. Cells of outer chlorenchyma had the lowest volume and thickest cell walls (1.2 μm), grading to water-storage parenchyma with highest volume and thinnest cell walls (0.6 μm). Water-storage parenchyma, which constituted 64% of leaf volume, accounted for 81% of tissue water capacitance and 72% of the plant total, with inner chlorenchyma having most of the remainder. Therefore, when soils are dry, water-storage parenchyma is the principal source of water for meeting transpirational demands and keeping chlorenchyma hydrated. This differs from all-cell succulent leaves, in which every cell shares equally in intraleaf water relations.

5.2.3 Windowed Leaves

Window-leaved plants of lowland deserts are succulents that mostly reside belowground (Marloth 1909; Schmucker 1931; Karsten 1931; Schanderl 1935; Seybold 1955; Rauh 1974; Hammer and Hartmann 1990). Only the flattened or domed leaf tips are exposed to the sun, and they have one large to numerous small transparent to translucent "windows" (*Grossfenster* versus *Kleinfenster*) and clear, water-storing central mesophyll ("water parenchyma"), which in combination function as a light shaft to transmit PPFD to subterranean assimilatory chlorenchyma (Fig. 5.9A,B). Chlorenchyma, with its associated stomata, is located in a mantle of the buried leaf, and CO_2 diffuses through stomata from the soil, not through astomatic windows.

This type of geophylly, wherein the leaf tip is at the soil surface or even completely covered part of the year, evolved convergently in southern Africa from succulent rosette plants in various genera of Aizoaceae as well as in

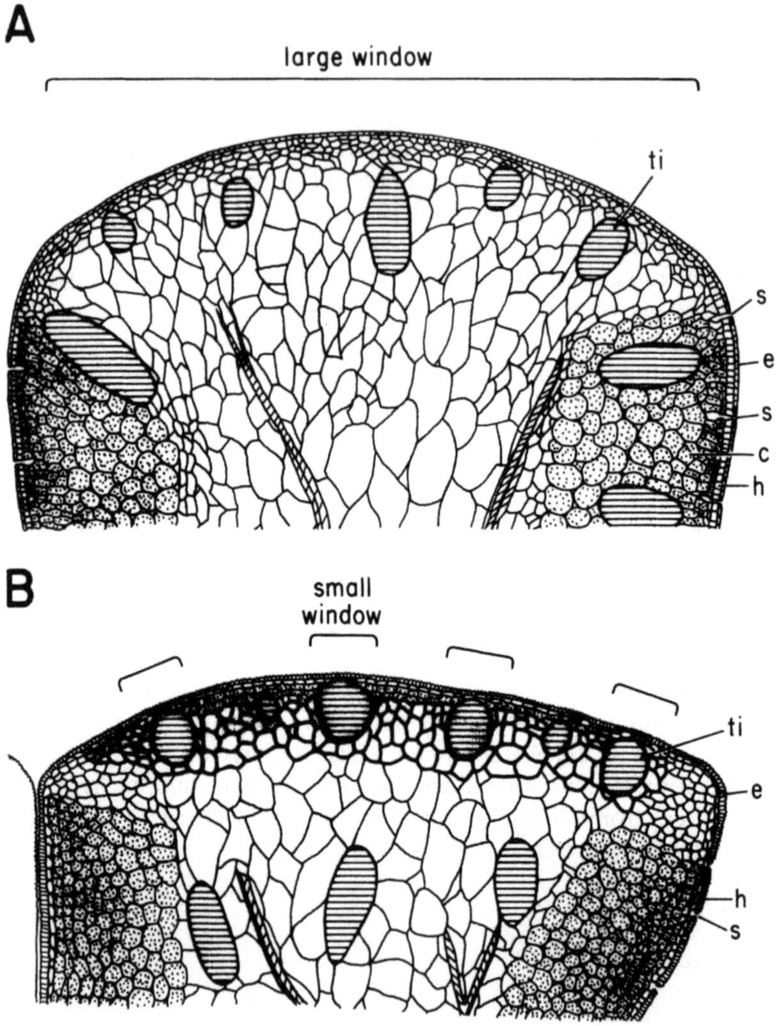

Fig. 5.9A,B. Windowed-leaved leaf succulents of Aizoaceae. **A** *Fenestraria aurantiaca*, in which an unpigmented leaf tip functions as a large translucent window (Grossfenster), permitting sunlight to penetrate the thick leaf, thereby illuminating peripheral chlorenchyma (*stipple*) of the subterranean leaf. **B** *Lithops fulviceps*, in which the leaf tip has some thick-walled, chlorophyll-bearing cells separated by translucent small windows (Kleinfenster); here also subterranean peripheral chlorenchyma is the chief photosynthetic tissue. In both cases, stomata (*s*) are present in epidermis (*e*) opposite peripheral chlorenchyma (*c*) and totally absent on the leaf tip. Large tannin idioblasts (*ti*) occur in translucent tissue as well as in chlorenchyma. Vacuolar betalain pigments are produced in hypodermis (*h*)

Haworthia and *Bulbine* (Asphodelaceae); in each case, during drought the plant is drawn deeper into the soil by contractile roots. Of these only certain Aizoaceae occur in the Namib, and here relatively few of the species. Large windows are best illustrated in the monotype *Fenestraria aurantiaca* (Fig. 5.9A), whereas small windows (as small as 25 μm in diameter), sometimes described as pellucid dots, characterize numerous species of *Conophytum* (Hammer and Hartmann 1990) and *Lithops* (Cole 1987) and contribute to their cryptic coloration. Probably little photosynthesis occurs in the scanty chlorenchyma at the leaf tip (Fig. 5.9B), because these cell walls are too thick to permit efficient CO_2 diffusion, and no stomata occur there.

Adaptive significance of windowed geophylls has been suggested as either preventing herbivory and damage from ungulate hooves or a mechanism to avoid any of several limiting abiotic factors. One popular interpretation has been that this plant design reduces leaf temperature and light interception, assuming that light levels are excessive for surface leaves and heating is deleterious for cell viability, transpiration, or photosynthesis. In Aizoaceae, old measurements of PPFD interception by assimilatory tissues, ranging from 10–60% surface radiation (Schanderl 1935; Seybold 1955), have been accepted as evidence that these leaves receive adequate PPFD (Hammer and Hartmann 1990), although many succulent plants are now known to be PPFD-limited even in full sun (Gibson and Nobel 1986; Nobel 1988, 1991b). A suggestion that high-intensity sunlight destroys chlorophyll (Rauh and Hutchison 1973; Rauh 1974) has not been tested in these species. Burial of leaves as a way to lower leaf temperature has been challenged by three recent studies, showing that in *Lithops* and *Haworthia* leaf temperature followed the diurnal course of soil temperature with little deviation but greatly exceeded midday air temperature and even stored more heat than the surrounding soil (Eller and Nipkow 1983; Eller and Grobbelaar 1986; Nobel 1989a). However, none of the measured leaf temperatures was close to lethal thermal limits of water-storing central mesophyll (55.3–61.0 °C) or chlorenchyma (66.4–68.7 °C; Nobel 1989a), although overheating and burning are recognized problems in growing small Aizoaceae (Hammer and Hartmann 1990).

Gas exchange measurements, including knowledge of CAM activity, on these species are too inadequate to analyze window-leaved species in terms of strategies to maximize CO_2 uptake, minimize transpiration, or increase WUE. Certainly total volume of assimilatory tissue is several times greater in subterranean leaves than it would be if the leaf tip was the sole photosynthetic region. Transpiration is prevented from the astomatic window and would be reduced because boundary layer is increased above a window positioned at soil level. Belowground, stomata open next to soil, which also provides a thick boundary layer, hypothetically lowering stomatal water vapor conductance and thereby affecting WUE.

5.2.4 *Epidermis of Aizoaceae*

Two basic types of epidermis have been described from Aizoaceae (Ihlenfeldt and Hartmann 1982; Ihlenfeldt 1983). The xeromorphic type has cells with thick outer periclinal walls and cuticle, sometimes with calcium oxalate sand crystals imbedded in S layers of the wall. In these, stomata are sunken the depth of the thickened cell wall. Species with xeromorphic epidermis are described as being "savers," minimizing water loss under all conditions while relying on centrally stored water, surviving extreme drought with persistent leaves (von Willert et al. 1980). Xeromorphic epidermis also tends to be more reflectant, caused probably by the imbedded crystals and resulting in lower transmission of PPFD and IR (Ihlenfeldt 1983). If universally present around the entire leaf, xeromorphic epidermis would function as a hydrostat. The old, dried leaf pair may serve as an additional barrier to water loss (Fearn 1977).

Other Aizoaceae have a composite epidermis with unspecialized, lightly cutinized groundmass cells and stomatal apparatus partially or totally concealed by bladder cells. A bladder cell is an extremely large, glistening, protruding structure having an unevenly thickened wall; it contains much water and accumulates high concentration of salts in the huge vacuole (Haberlandt 1914). As much as 50% of leaf water can be stored in bladder cells for rehydrating chlorenchyma. Cuticle is multilayered (Kramer 1979), almost impermeable to water and solutes (Lüttge and Smith 1984). These cells undoubtedly affect solar radiation levels received by leaf chlorenchyma and heat transfer within the boundary layer, but results have been inconclusive (Tanner and Eller 1986). Under water stress, bladder cell idioblasts lose turgor and collapse, leading to speculation that this significantly affects stomatal activity (Ihlenfeldt 1983) or lowers boundary layer conductance.

Aizoaceae with bladder cells have been judged as "opportunists," having higher water turnover, higher growth rates, but poorer adaptations for surviving drought (von Willert et al. 1980). For these CAM species it has been suggested that bladder cells may add in nocturnal opening of stomata under foggy conditions, with water influx through epidermis because of the reverse water potential gradient (Ihlenfeldt 1983). Species with prominent bladder cells occur in several desert regions, most abundant in southern Africa but also for annuals of *Mesembryanthemum* in the northeastern Sahara-Arabian region and *Tetragonia maritima* in foggy Atacama. Bladder cells also occur on succulent lamina of certain species of Crassulaceae and petioles of *Oxalis*.

5.2.5 Areoles and Spines of Cacti

The spine-bearing areole of Cactaceae is an axillary bud with sylleptic, determinant development, forming sclerified primordia as spines that elongate via an intercalary meristem (Gibson and Nobel 1986). An areole typically has a dense cover of uniseriate trichomes. This structure evolved from ancestral forms that had short shoots with photosynthetic leaves rather than spines (Buxbaum 1950). In the desert Cactaceae, succulent stems are the chief photosynthetic organ and leaves are vestigial and inconspicuous, eliminated via natural selection because leaves would block PPFD from reaching stem chlorenchyma and thereby lower primary productivity.

Traditional explanations for spines on plants focused on their function in defense against herbivores. This function fits observations that few mammalian herbivores preferentially consume stem succulents that are protected by spines.

In deserts, where spine densities and forms are diverse, researchers have shown that spine and trichome characters influence plant temperature, PPFD interception, and shoot water relations (Nobel 1978b, 1980a,b,c, 1983b; Gibson and Nobel 1986). Using a computer thermal exchange model (Lewis and Nobel 1977) and field measurement on *Ferocactus acanthodes* and several species of columnar cacti of the Sonoran Desert, Nobel studied the importance of spines and thick trichome mats on apical temperatures. He determined that dense apical vestiture had a significant insulating effect by lowering daytime maximum apical temperature and raising nighttime minimum apical temperature, while relatively dense spination had a similar effect (Fig. 5.10). The combined effect of pubescence and spination appeared to determine freezing avoidance by columnar and barrel cacti along a north-south gradient.

Spines, which provide some cold-night protection for cacti, and presumably spiny species of *Euphorbia*, carry some costs, in addition carbon and energy costs to make them. When spines were removed from the desert cylindropuntia *Opuntia bigelovii*, receiving PPFD of 9 mol m^{-2} d^{-1}, stem volume increased 60% over 2.5 years (Nobel 1983b). It can be shown that spine shading limits nocturnal acid accumulation in desert cacti growing even in full

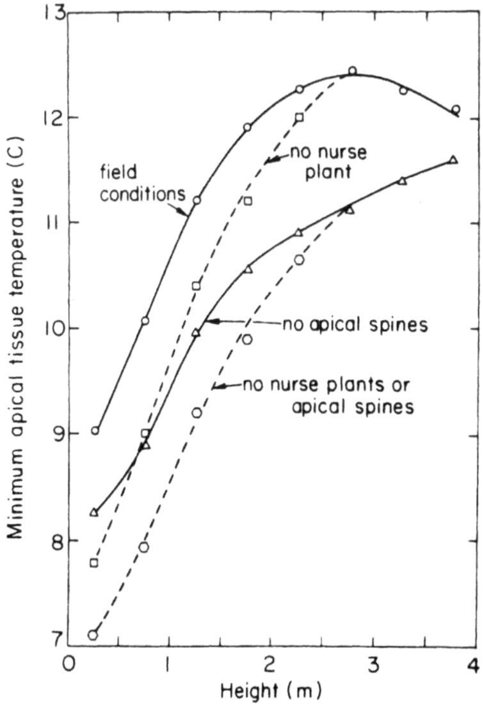

Fig. 5.10. Influence of spination on shoot tip and nurse plant cover on nocturnal minimum apical temperature. (After Nobel 1980b)

sun, e.g., a full-sun barrel cactus having spines that block about 50% of incoming PPFD (Gibson and Nobel 1986).

Presence of spines along a succulent stem has been credited with creating a thicker boundary layer and thereby reducing transpiration, but this effect is actually very small because stomatal water vapor conductance dominates this gas diffusion pathway (Gibson and Nobel 1986). However, spines may affect the longer CO_2 diffusion pathway by influencing turbulence and boundary layer on stems in ways that have not yet been modeled.

5.2.6 Euphorbiaceae with CAM Stem Photosynthesis

Species of *Euphorbia* in the Old World and *Pedilanthus* in North America have CAM within aphyllous succulent stems, including forms with tuberculate and ribbed stems that are convergent with Cactaceae. Relatively little physiological and anatomical research has been conducted on these species. *Euphorbia caducifolia* is a CAM species that is leafless most of the year and depends primarily on photosynthesis of evergreen cylindrical stems (Sen 1968; Sen et al. 1971; Sen and Chawan 1972). Leaves have relatively thin cuticle and two to four times greater stomatal densities, but are shed at high leaf water potential, while stem water content is still very high.

5.2.7 Stapelieae

The tribe Stapelieae has radiated in arid and semiarid eastern and southern Africa and Arabia (Plowes 1990). These are low caespitose to simple and unbranched (*Pseudolithos*) CAM stem succulents, evolved from a stem succulent with well-developed deciduous leaves, represented by *Frerea indica* of India, which has CAM stems and C_3 leaves (Lange and Zuber 1977, 1980). Leaf blade evolved to be a minute caducous structure, and the lower leaf and stem form a tubercle, barely discernible, prominent like thorns, or coalesced vertically into ribs. Many stapeliad genera have representative warm desert species (Table 5.1).

Stem anatomy of desert taxa from numerous genera is remarkably similar, suggesting that the tribe is relatively uniform. A smooth, relatively thin, transparent cuticle, lacking epicuticular wax, coats the stems, so that epidermal wall with cuticle rarely exceeds 5 µm in thickness, and these tabular cells have a thin inner periclinal walls (2–4 µm). Ability of these plants to survive in some of the harshest desert environments clearly is not attributable to obvious structural barriers to water loss. Stomata are superficial, if sunken only a few micrometers, and generally occur at extremely low densities (Zemke 1939). There is no hypodermis. Cortical chlorenchyma, which typically extends 1 mm or more to the vascular cylinder, is composed almost exclusively of isodiametric parenchyma; cell walls characteristically are at least 1 μm thick and may be 5–8 μm thick, but still unlignified, in those with prominent tubercles (*Caralluma* and *Huernia*). Fibers of outer primary phloem are few, thin-walled, and short with rounded tips. Inner cortex and pith often store abundant starch.

The only field ecophysiological study on Stapelieae was performed on *Caralluma negevensis*, an arido-active cushion-like CAM species of the Negev, occurring among rocks and in shady habitats where water runoff collects (Lange et al. 1975b). In early spring, following winter rains, stems were fully turgid and experienced nocturnal CO_2 uptake, at low rates with maximum values before midnight, but as nighttime temperatures increased above 20 °C, rate of nocturnal CO_2 uptake sharply declined. During daylight, substantial amounts of CO_2 from decarboxylated malate were released, presumably because epidermis is weakly cutinized, causing this plant to exhibit carbon loss during the entire hot drought. Thick walls in chlorenchyma would be expected to lower liquid-phase CO_2 diffusion rates and may contribute to the slow growth rates of these species.

5.2.8 Leaves of Crassulaceae

Crassulaceae occur in many semiarid habitats worldwide, especially where they root in rocky outcrops. In North America, only a few species can be found on the margins of fog desert (*Dudleya*), forming basal rosettes of thick leaves. In winter rainfall zones of southern Africa, several genera of Crassulaceae have speciated as drought-deciduous and evergreen perennials and succulent to fleshy-leaved annuals. On both continents, leaves (CAM in desert species) are models of simplicity, consisting of unremarkable epidermis and thin-walled mesophyll parenchyma and lacking sclerenchyma. In woody species of *Tylecodon*, leaves abscise during its reproductive phase in spring and then are renewed in late summer (von Willert et al. 1992).

Stems of Crassulaceae are not photosynthetic and, when persistent, are soon covered with periderm.

5.2.9 Fouquieriaceae

The family Fouquieriaceae consists of 11 species of C_3 stem-succulent, divaricate shrubs (ocotillos), woody small trees, and short to tall monopodial pachycauls, all with long shoot-short shoot organization and so-called mesomorphic leaves (Henrickson 1969a,b, 1972, 1975, 1977; Humphrey 1974). All species are native to arid or semiarid habitats of Mexico and the southwestern United States, but species in the driest habitats tend to possess the most extreme structural modifications.

Each long shoot bears rigid, helically alternate spines, presumably an adaptation from the origin of the family in thorn scrub and thorn forest of Mexico, The "fouquieraceous spine" forms from the fibrous abaxial portion of a long-shoot leaf petiole via a horizontal abscission layer (Fig. 5.11A), thereby shedding the adaxial petiole and elliptical lamina (Robinson 1904; Humphrey 1931, 1935; Henrickson 1969c, 1972, 1977). Spine fibers are an extension of stem cortical sclerenchyma, underlying a decurrent stem ridge (Fig. 5.11B). In *F. splendens* and the gypsophilous *F. shrevei*, as stem diameter increases with secondary growth, sclerified stem ridges become separated laterally to form a network of furrows, covered by layered translucent periderm, which permits PPFD transmission to photosynthetically competent chlorenchyma of the outer cortex. Desert researchers suspected that such green stems are assimilatory, but in *F. splendens*, uptake of exogenous CO_2 has never been demonstrated because no gas exchange pathway occurs through the impermeable furrow cork layers; however, its stem reassimilates endogenous CO_2, an adaptation to refix respiratory CO_2 (Mooney and Strain 1964; Nedoff et al. 1985). *Fouquieria shrevei* is unique for the family in having resin-like deposits of waxes on furrow periderm (Henrickson 1972).

Green to yellow bark of the pachycaulous boojum, *F. columnaris*, consists of thin, suberized periderm covering cortical chlorenchyma, but shows no significant CO_2 uptake under conditions of either high or low soil moisture (Franco-Vizcaíno et al. 1990). Cortical chlorenchyma (to 700 μm) contains sclereid nests, giving the surface a rock-like texture that appears to deter rodent herbivory (Henrickson 1969b, 1969c, 1972). A thick layer of sclereids occurs beneath cortical chlorenchyma.

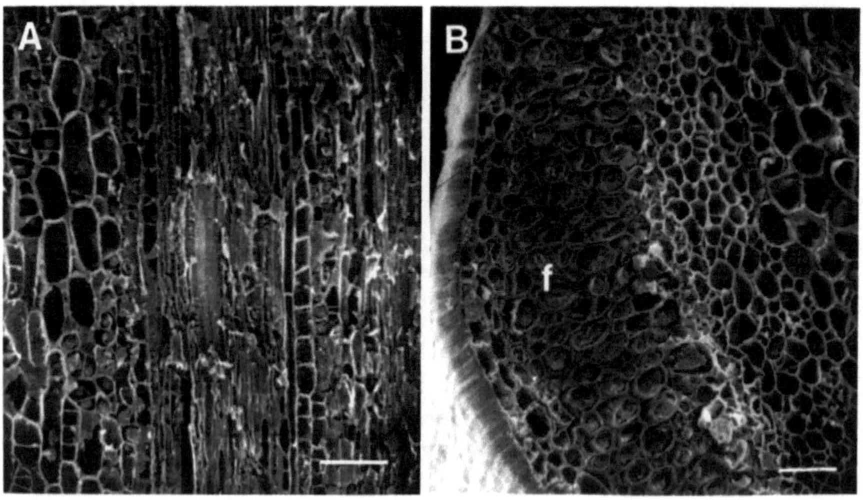

Fig. 5.11A,B. *Fouquieria burragei.* **A** Longisection through young lower leaf when abscission layer is forming between adaxial petiole and lamina. **B** Stem cortical sclerenchyma fibers (*f*) that provides hardness to decurrent stem ridge. *Bar* 50 μm

All species in the inner cortex possess an anastomosing network of thin-walled water storage tissue, containing abundant reducing sugars and oil globules (Scott 1932; Henrickson 1969c). Nodes of this network are opposite short shoots, suggesting that the network is responsible either for delivering nutrients to emerging leaves or for transporting photosynthates from mature ones (Henrickson 1969c).

Fouquieria splendens and *F. shrevei* are leafless most of the year, but leaves appear shortly after each brief warm-season rainfall in fascicles on short shoots, often fully formed within 48 h (Darrow 1943). These species are drought-deciduous with water stress, leaving immature primordia up to 1.5 mm in the axillary bud, but may be stimulated to form a flush of leaves with each rainfall. Leaves of *F. columnaris* in the field can persist for more than 6 months, even during progressive drought, and still show active transpiration with low CO_2 uptake (Franco-Vizcaíno et al. 1990).

Laminae of all desert species of *Fouquieria* are relatively narrow (mostly 3–14 mm) and amphistomatic and have isolateral mesophyll (Fig. 5.12A,B). *Fouquieria columnaris* tends to have much spongy mesophyll (Henrickson 1972). Ratio of stomatal densities on abaxial versus adaxial leaf surfaces was unity for *F. shrevei* but higher for *F. splendens*, which had the thin leaves and highest percentage of palisade parenchyma, and lower for *F. columnaris*, which had the thickest leaves (to 0.5 mm) and low percentage of palisade parenchyma.

Succulence is highly developed in the main stem of *F. columnaris*, which may achieve a basal diameter of 0.7 m (Henrickson 1969b,c). Boojum has

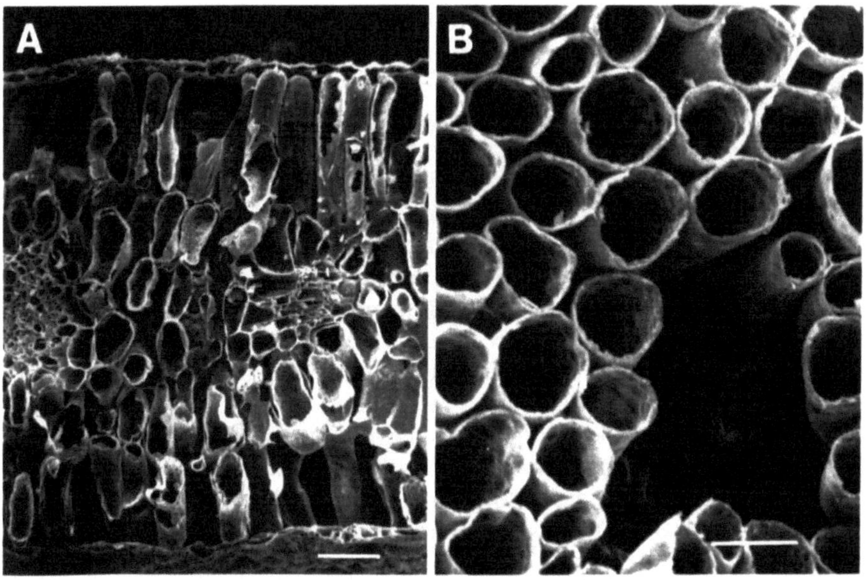

Fig. 5.12A,B. Leaf anatomy of *Fouquieria columnaris*. **A** Transection of amphistomatic leaf with isolateral mesophyll. **B** Paradermal view through adaxial palisade mesophyll, showing low degree of cell touching and abundant intercellular air space and large substomatal chamber. *Bar* 25 μm

extensive parenchymatization of secondary xylem and a wide water-storing pith, contributing to high water capacitance of old stems (3.6 kg H_2O m^{-3}; Nilsen et al. 1990). Leaf cells have higher osmotic potentials and stiffer walls than nonsucculent leaves, so they lose turgor rapidly with an increase in water deficit. In *F. columnaris*, leaves are buffered against wilting by high water capacitance of the trunk, and experience extremely small changes in leaf water potential over the course of the day.

Ocotillos have been compared with shrubby and arborescent Didiereaceae of semiarid Madagascar, as an example of convergence in that both have wand-like shoots and ephemeral short-shoot leaves. In contrast, didiereas are CAM plants (Kluge and Ting 1978) and have greater development of water-storing tissues in the cortex and pith (Rauh 1961).

5.2.10 Elephant Trees, Pachycauls, and Caudiciforms

Similar to *F. columnaris* are so-called elephant trees, such as *Pachycormus discolor* (Gibson 1981) and *Bursera microphylla*, having pachycaulous stems with translucid exfoliating bark that covers water-storing parenchymatous tissue (Franco-Vizcaíno et al. 1990). As with *F. columnaris*, nonsucculent C_3 leaves are the principal photosynthetic organs, where exogenous CO_2 assimilation occurs, but bark photosynthesis is critical during leafless periods to maintain adequate carbohydrates, especially by refixing the bulk of respiratory CO_2 within the stem. Trunk water volume is high while leaf area is small, therefore during early drought these plants are buffered against large daily leaf water potentials (Nilsen et al. 1990).

Stems of *Pachycormus* contain remarkably light-weight wood, consisting mostly of very wide, nucleate libriform fibers with very thin, unlignified cell walls but filled with water, which provides turgor pressure sufficient to make the thick limbs solid and strong (Gibson 1981). Wood fibers are also packed with starch grains, which probably contribute matrix forces for retaining the water. Very thick bark is moisture-rich and considered also to be an organ for water storage. Thick-barked desert Burseraceae (*Bursera*, *Boswellia*, and *Commiphora*) have a larger component of cell wall materials in woods. Stems and leaves of these elephant trees have abundant secondary compounds to deter predators that would utilize stem resources.

It is assumed, but not yet determined experimentally, that stem photosynthesis of most caudiciforms and pachycauls contributes little, if any, annual assimilation of exogenous CO_2. However, a study of the caudiciform *Adenium obesum* showed substantial stem photosynthesis when leaves were absent (Stocker 1971). Water storage tends to be present throughout a stem, with no clearly identifiable tissue where water storage is the sole function.

In *Pachypodium namaquanum*, some wide, velvety leaves of the apical canopy persist during dormancy (Rowley 1987). Its leaves are amphistomatic, with superficial stomata and thin cuticle (Lee 1912). Authors have assumed that large leaves of these succulent-stemmed species are "mesomorphic". Indeed, species have bifacial leaves, e.g., *Dorstenia gigas* (Fig. 5.13A), *Dendrosicyos socotrana*, and species of *Adenia*. *Pachycormus discolor* has hypostomatic, bifacial microphyllous leaflets in which a single adaxial palisade layers accumulates abundant tannins (Fig. 5.13B; Gibson 1981). Nonetheless, the majority of microphyllous species studied to date have amphistomatic leaves with isolateral mesophyll, like those of nonsucculent desert shrubs.

5.2.11 Asteraceae

Senecio is a cosmopolitan C_3 genus with a clade, often segregated as the genus *Kleinia*, for CAM stem succulents of arid and semiarid Old World. One of the true deserticolous species, *S. deflersii* of South Yemen, has scale-like, C_3 hypostomatic, bifacial leaves but thick photosynthetic CAM stems covered by a multiseriate epidermis with low stomatal density (17.2 mm^{-2}) and slightly sunken stomata over isodiametric cortical chlorenchyma to 1 mm thick (Fioretto and Alfani 1988).

5.3 Fleshy Leaves

Fleshy leaves typically are shed during water stress but may persist longer than nonsucculent leaves of sympatric drought-deciduous shrubs and subshrubs. Among glycophytes are deserticolous fleshy-leaved species of *Lycium* and

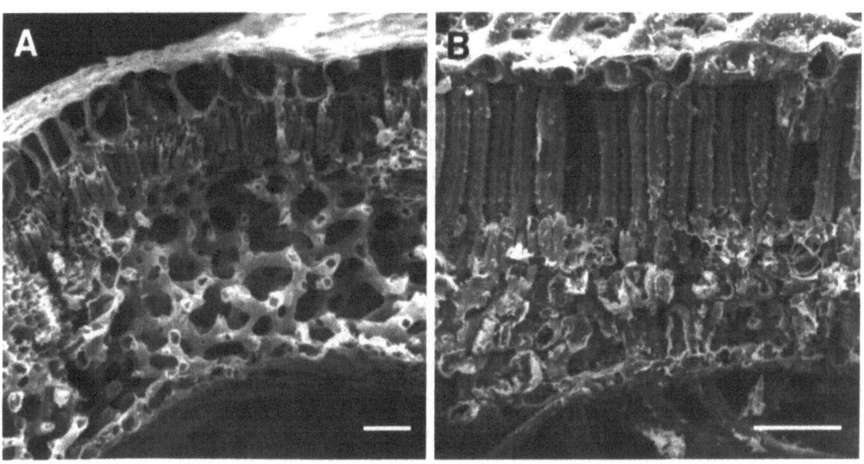

Fig. 5.13A,B. Transections of bifacial leaves from desert pachycauls. **A** *Dorstenia gigas*. **B** *Pachycormus discolor*. *Bar* 50 μm

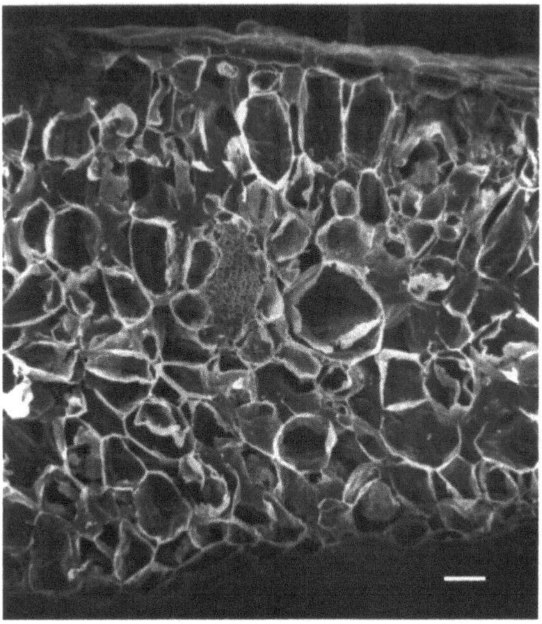

Fig. 5.14. *Lycium shawii*; leaf transection without palisade parenchyma. *Bar* 50 μm

Zygophyllum. *Heliotropium curassavicum* is a fleshy herbaceous perennial that is somewhat salt-tolerant and shows how this group of desert plants is transitional between lowland desert and saline habitats. Species of *Pelargonium* in southern Africa are transitional with succulents.

Thick leaves of *Lycium*, found on all continents, have large mesophyll parenchyma cells and seldom exhibit well-defined palisade (Fig. 5.14).

Species of *Zygophyllum* display an interesting range of growth habits and phenologies. Included are C_3 fleshy-leaved and sclerophyllous evergreens, drought-deciduous phanerophytes and chamaephytes, and therophytes, including *Zygophyllum simplex*, which is a C_4 leaf-succulent annual (Ziegler et al. 1981; Ismail 1983). Species of *Zygophyllum* can have shoot water content as low as 27%. Among the chamaephytes *Z. dumosum*, *Z. coccineum*, *Z. decumbens*, and *Z. qatarense*, fleshy leaflets abscise during the dry season, but the petiole persists as a photosynthetic organ (Evenari 1938b; El-Monayeri et al. 1981; Ismail 1983), and thereafter petiolar epidermis undergoes periclinal divisions to form a multiple epidermis with sunken stomata (Evenari et al. 1971). Succulence ratios were 4.8–7.4 in green and yellow leaflets or petioles of Egyptian species of *Zygophyllum* (Fahmy et al. 1990).

Fleshy leaves of *Zygophyllum*, as illustrated by *Z. dumosum*, have modest cuticle and palisade parenchyma that is hardly distinguishable from water-storing central mesophyll cells (Volkens 1887; Evenari 1938b; Stocker 1972). Stomata may be either superficial or somewhat sunken.

6 Special Topics in Water Relations

Desert researchers have attempted to identify plant adaptations for either collecting or conserving water. Relatively few structural subjects have been pursued with sufficient experimentation to draw generalizations that fit desert plants. The following are particularly interesting topics on water uptake, transport, and management that have been investigated.

6.1 Xerohalophytes

Inland saline habitats have formed within high-evaporation basins and lowlands of the desert topography. On the one hand, these habitats temporally may have substantial soil water available for plant growth, but such soils contain many salts; especially NaCl is dominant. The electrolytes Na^+ and Cl^- are extremely toxic to most plants at relatively low soil water concentrations, due to deleterious effects on cellular metabolism and ultrastructure. True halophytes, which thrive when given water having greater than 0.5% NaCl, have evolved convergently in numerous families by possessing structural, phenological, physiological, and biochemical mechanisms for salt resistance (Waisel 1972; Poljakoff-Mayber and Gale 1975; Flowers et al. 1977; Osmond et al. 1980; Sen and Rajpurohit 1982; Munns et al. 1983; Staples and Toenniessen 1984; Cheeseman 1988). Xerohalophytes, the desert representatives, and coastal halophytes share these mechanisms, and in most cases belong to the same phylogenetic lineages, and detailed analysis is still needed to determine where halophytism evolved in each family occupying both habitats.

Using terminology of Levitt (1972), salt resistance can involve salt tolerance or salt avoidance. Salt tolerance involves physiological and biochemical adaptations for maintaining protoplasmic viability as cells accumulate electrolytes and an osmotic adjustment to extract water from soils with low water potentials, whereas salt avoidance involves structural with physiological adaptations to minimize salt concentrations in cytosol and cell organelles or physiological exclusion by root membranes (Caldwell 1974). Considering only structure-function adaptations, organisms have been classified as excretives, with glandular cells capable of secreting excess salts from plant organs, versus succulents, increasing water content within large vacuoles to minimize salt toxicity, or excluders versus includers, respectively (Staples and Toeniessen 1984).

6.1.1 Vesiculated Trichomes of Atriplex

The genus *Atriplex* includes xerohalophytes present in every warm desert as well as species that tend to be glycophytes but tolerate some soil salinity. Leaves of all species are concealed by vesiculated trichomes (Fig. 6.1), consisting of a stalk with one to several parenchyma cells having dense cytoplasm and a terminal bladder cell, 80–200 μm in width, with a huge vacuole where salts are concentrated. These species are often classified as excretives, but actually release salt only when they rupture or collapse. Either dense or ruptured trichomes can increase leaf reflectance and thereby significantly affect the energy budget (Pearcy et al. 1974).

Several excellent reviews in detail describe the structure and known cell biology of these salt-accumulating trichomes (Thomson et al. 1969; Thomson 1975; Osmond et al. 1980; Schirmer and Breckle 1982; Bennert and Schmidt 1983). Briefly summarized, mesophyll cells maintain relatively low concentrations of Na^+, K^+, and Cl^- because excess cations and anions are transported via epidermis and stalk cells into vacuoles of bladder cells (Osmond et al. 1969; Pallaghy 1970). The gland is covered by cuticle, although lateral walls of stalk cells may not be fully cuticularized (Smaoui 1971; Thomson 1975).

Vesiculated trichomes of *Atriplex* also are particularly effective in attenuating UV-β transmittance (Robberecht and Caldwell 1978). Hypothesized other functions of these structures were proposed before a role in salt resistance was understood.

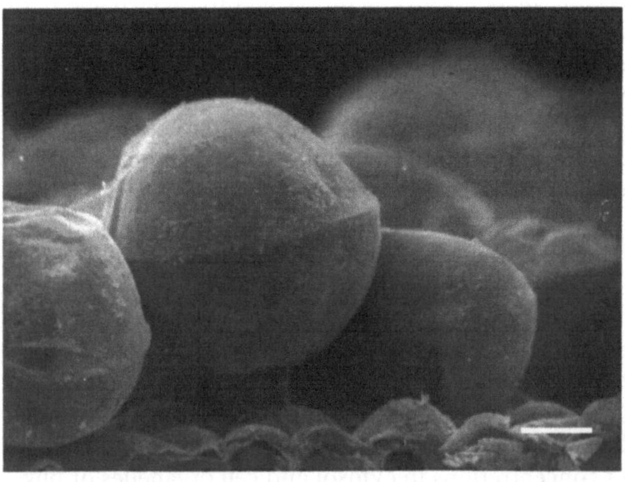

Fig. 6.1. Vesiculated trichome on a leaf of *Atriplex canescens*. *Bar* 10 μm

6.1.2 Excreting Salt Glands

Excreting salt glands occur in tribes of grasses, Chlorideae, Sporoboleae, and Aeluroideae, including the halophytes *Distichlis spicata* and *Aeluropus litoralis*, which may occur in desert alkali sink, and several glycophytes or facultative halophytes of other genera (Liphschitz and Waisel 1974, 1982). Each gland (Fig. 6.2A) is a bicellular structure consisting of a cutinized cap cell with dense cytoplasm and an uncutinized basal cell (Thomson 1975; Oross et al. 1985). A collecting chamber occurs between cuticle and wall of the cap cell, where salt solution accumulates and probably is forced through tiny pores when cuticle is stretched as hydrostatic pressures builds.

Among xerohalophytic excretives of warm and cool deserts are *Tamarix* and *Reaumuria* (Tamaricaceae) and *Frankenia* (Frankeniaceae). A gland (Fig. 6.2B,C) consists of six upper secretory cells and two highly vacuolate subbasal collecting cells (Shimony and Fahn 1968; Liphschitz and Waisel 1982; Bosabalidis and Thomson 1984; Fahn 1988). A collecting compartment forms between secretory

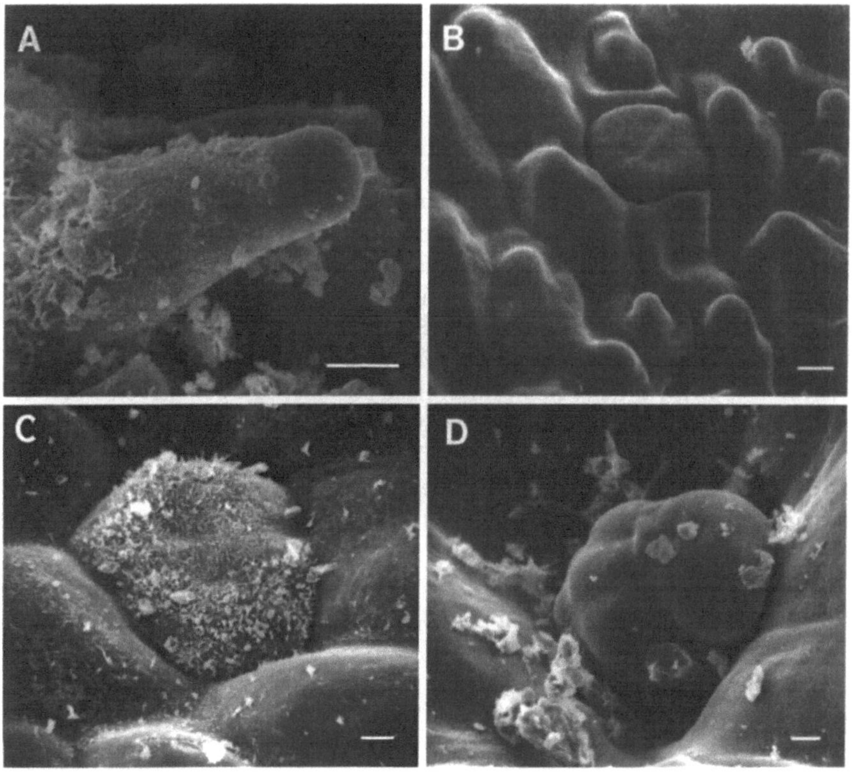

Fig. 6.2A–D. Types of salt glands on desert xerohalophytes. **A** *Distichlis spicata*. **B** *Tamarix ramosissima*. **C** *Frankenia palmeri*. **D** *Nolana mollis*. *Bar* 5 μm

cells and the outer cuticular shell, and passage of salts probably occurs from there via cuticular pores (Thomson 1975). Highly reduced leaves of *Tamarix*, which are appressed to the stem, and revolute leaves of *Frankenia* are two notable examples of hypostomaty in deserts (Weiglin and Winter 1991).

6.1.3 Halophytic Succulence

Succulence, also called fleshiness, is strongly expressed in many xerohalophytes from playas or in alkali sink. Such desert representatives are *Suaeda, Zygophyllum* (especially therophytes), *Nitraria, Mesembryanthemum, Aizoon, Halophytum, Trianthema,* and *Heliotropium*, and stem-succulent Chenopodiaceae, e.g., *Salicornia, Arthrocnemum,* and *Allenrolfea,* in which photosynthetic mesophyll from highly reduced foliar primordia is continuous with stem cortex (James and Kyhos 1961; SaadEddin and Doddema 1986), and *Haloxylon.*

Succulent leaves typically are amphistomatic and have a number of distinctive features, which distinguish them from typical nonsucculent desert leaves or are expressed when xerohalophytes are grown under saline and nonsaline conditions (Jennings 1968, 1976; Waisel 1972): (1) thicker leaves, hence higher *V*/*A*; (2) larger parenchyma cells in most or all tissues; (3) lower stomatal densities, caused by having greater expansion of groundmass epidermal cells; (4) smaller intercellular air spaces; (5) lower chlorophyll content per leaf surface area; and (6) higher cell wall elasticity. These structural changes permit greater water storage, hence greater accumulation of ions in vacuoles. Salts are eliminated from the plant when leaves or shoots are shed (Albert 1975). In Chenopodiaceae with articulated shoots, photosynthetic tissues abscise following water stress (Fahn 1963), presumably also eliminating salts. Aizoaceae sequester Na^+ and Cl^- in bladder cells (Steudle et al. 1975; Kramer 1979), and relatively thin-leaved desert species of *Mesembryanthemum* shift from C_3 photosynthesis to CAM and develop pronounced succulence when placed under either saline or water stress conditions (Winter 1973; Winter and Troughton 1978).

Species of *Atriplex*, which are salt-excluders, also produce thicker, fleshier leaves (C_4) when grown under the most extreme tolerable saline conditions for that species (Black 1958). From different sites, leaves of a species can vary greatly in fresh weight to leaf area, from almost nonsucculent to very fleshy.

In *Prosopis farcta*, sodium is selectively extracted from xylem sap and retained in stem tissues to avoid salt damage to leaves (Waisel 1972).

6.1.4 *Nolana mollis* and Other Atacaman Shrubs

Nolana mollis is a succulent-leaved shrub of the fog belt in the Atacama. On a leaf NaCl-excreting trichomes occur in depressions (Fig. 6.2D), part of a network of troughs. In situ experiments demonstrated that under conditions of high humidity leaves with surface salt can quickly accumulate substantial amounts of water, which drips off and wets the soil beneath the shrub canopy, whereas desalted leaves condense no surface water (Mooney et al. 1980a).

Nolana mollis and convergent, sympatric shrubs accumulate high levels of salts in their fleshy leaves (Rundel et al. 1980). Ash content of *N. mollis* (34% dry weight), *Heliotropium pycnophyllum* (45%), *Tetragonia maritima* (34%), and *Eremocharis fruticosa* (23%) were remarkably high, higher than most succulent halophytes, even though soils of the Atacaman community are not saline. For *N. mollis* and *T. maritima*, NaCl content was 18 and 15% dry weight, respectively. Each shrub had an ash pile beneath its canopy from gradual decomposition of abscised leaves.

Mesophyll of these leaves consists mostly of large, thin-walled, isodiametric parenchyma, usually with relatively little or no palisade development, and mesophyll may contain crystals in every cell.

6.1.5 Crystal Deposition in Shoots

Although presence of crystals in photosynthetic organs has been regarded as a xeromorphic feature (Pyykko 1966), occurring in one-third of all species sampled (Fahn and Cutler 1992), in general, crystals tend to be abundant in particular desert plants, e.g., hypodermal druses of *Opuntia*, mesophyll raphides in leaves of Agavaceae and Aizoaceae, and prismatics and druses in parenchyma of Fabaceae, but relatively sparse or absent in nonsucculent leaves of many others. Leaves of evergreen shrubs and fleshy or succulent structures tend to have abundant crystals. Crystals in stem tissues and petioles have not been carefully surveyed.

6.2 Water Uptake

6.2.1 Roots

6.2.1.1 Morphology and Depth

Soil water potentials in typical desert lowlands may remain below −10 MPa during drought months, often corresponding to periods when daytime upper soil temperatures are very high, sometimes exceeding 70 °C.

When upper soil is dry, some perennials maintain favorable shoot water status by having deep roots that yield hydraulic lift from deep wet soil or gravel layers. Dominant desert shrubs with relatively deep roots as well as horizontal superficial roots are *Larrea tridentata* in North America and *Zygophyllum decumbens* in the northeastern Sahara, extracting water from a comparatively large soil volume, but they grow with equally successful chamaephytes and low phanerophytes that have fairly compact, shallow roots (Cannon 1911; Evenari 1938a). In arid Australia, *Atriplex vesicaria* is a common shallow-rooted C_4 species, less than 0.3 m, growing in soil that infiltrates poorly, whereas *Maireana sedifolia* is a deep-rooted C_4 species in a soil with water penetration mostly 0.6–2.0 m (Carrodus and Specht 1965) and roots of *Atriplex nummularia* can reach 3.5 m (Sharma 1982). Differences in rooting depth of dominant coexisting shrubs have been observed in other deserts (Cody 1978; Manning and Barbour 1988; Rundel and Gibson 1996).

Many roots for desert species are located within the upper 0.20 m of soil, often with mean root depth 0.07–0.12 m, where plants can effectively collect any rain or fog drip that wets the upper soil profile, but far enough below the surface to avoid lethal high temperatures (Jordan and Nobel 1984; Nobel 1991c; Rundel and Nobel 1991). Shallow, horizontal roots commonly extend long distances from the canopy. One extreme is a Saharan dune bunchgrass, *Stipagrostis pungens*, having thin, unbranched, cord-like roots that spread at least 20 m from the plant base (Price 1911). In characteristic lowland desert scrub, shallow roots collect water in bare soil between shrubs (Cannon 1911), but shallow rooting also is common beneath nurse plants (Nobel 1989b). Subterranean rocks affect soil water content, influencing desert root proliferation and branching in wetter soil under rocks (Evenari et al. 1971; Nobel et al. 1992b). Nonetheless, in a region of the Atacama where the typical year is rainless, roots of the cactus *Copiapoa cinerea* were located mostly within 0.03 m of the surface, and no root exceeded 0.08 m (Gulmon et al. 1979). Likewise, in the arid Karoo, the extremely common Aizoaceae are very shallow-rooted, to take advantage of the unpredictable rains. In Death Valley, where the top 0.20 m of soil is dry during drought, *Tidestromia oblongifolia*, which is an arido-active herbaceous perennial, had its roots in the zone with most moisture, 0.25–0.40 m (Bennert and Mooney 1979). Mean root depth is lower for plants growing in sandy soils, which generally enables dune plants to maintain less negative water potentials than those growing off the dunes (Pavlik 1980).

Some desert phreatophytes exploit soil water at two levels (Huber 1935). In *Prosopis glandulosa*, maximum root proliferation occurred at 0.25 and 4.0 m depths, thereby utilizing rainwater and deep water, respectively, and enabling this species to maintain maximum leaf area during summer drought (Sharifi et al. 1983), but midday stem water potentials can be lower than −4 MPa while water loss is not curtailed (Nilsen et al. 1981). *Prosopis* in Death Valley utilizes lenses of freshwater at 9–26 m (Went 1955). *Prosopis farcta*, a halophyte, has well-developed rhizomes as well as deep roots to 15 m (Waisel 1972). Green-stemmed shrubs, which have deep roots and most often occur along wadis and runnels, may also have roots in upper soil layers; in *Retama raetam*, shallow, horizontal roots have significantly wider vessels versus those of deep, vertical roots (Fahn 1964; Fahn and Cutler 1992).

Perennial roots of summer-active desert Cucurbitaceae are often thick, few-branched taproots, concealed below the hottest soil surface and having most root tissue within the upper 0.50 m of soil (Dittmer and Talley 1964), which during summer drought may be extremely dry. Dittmer and Talley concluded that *Cucurbita palmata, C. digitata,* and *Apodanthera undulata* of North American deserts collected rainwater in shallow soils and relied on root capacitance for most shoot transpiration during summer drought. However, recent ecophysiological studies determined that root capacitance cannot provide enough water to match high transpiration rates, therefore the majority of water is drawn from deep soils using the few fine vertical roots (North 1992). In the Sahara, *Citrullus colocynthis* possesses a deep taproot and a fibrous lateral root system just above the water table to meet its high daily transpirational losses (Althawadi and Grace 1986). Presence of thick root periderm on taproots of desert cucurbits and *Convolvulus lanatus* (Dittmer and Roser 1963; Danin 1983) may protect these from desiccation, but certainly will slow water uptake through suberized cell walls.

Roots of annuals tend to be shallow, and are typically present during the rainy season, hence have this seasonal access to substantial soil moisture (Rundel and Nobel 1991). Among winter annuals, dicotyledons most often form a thin taproot to supply the basal rosette of leaves; they have low root/shoot ratios, and aboveground biomass increases for individuals with deeper, thicker taproots, which can extract soil moisture for longer periods and supply a larger reproductive shoot. Annual grasses, e.g., species of *Schismus* and *Bromus*, have greater absorbing surface on very shallow fibrous root systems, which appear to give these species competitive advantage in water uptake from the uppermost soil.

Although roots of some warm desert perennials tend to be shallow and long, root/shoot ratios are 0.5–1.3 for many species, and substantially lower for succulents (Rundel and Nobel 1991; Nobel 1994b). This contradicts earlier speculation that desert plants should have high root/shoot ratios to increase water absorption capacity (Volkens 1887). For example, in the highly productive C_4 bunchgrass *Pleuraphis rigida*, root dry weight was only 21% of total plant even though total root length exceeded 300 m for large, shallow-rooted individuals (Nobel and Franco 1986). On the other hand, absorbing surface area of roots has been shown to be several times greater than shoot surface (Nobel 1988).

6.2.1.2 Structural Responses to Drought and Rewetting

Two generalizations about desert plants have been that thin roots are abscised entering drought, and "rain roots" are formed very rapidly following a rain that effectively wets the upper soil. Statistics on occurrence of either are lacking, but recent studies have documented for *Agave deserti* shedding of 87% of lateral roots by 35 days of drought, rapid elongation of existing lateral within several hours after watering, and newly formed rain roots with mean lengths of 19 mm just 4 days after watering. In contrast, the sympatric *Ferocactus acanthodes* had only 12% shedding and no perceivable rapid root growth following rewatering (Nobel et al. 1990; Nobel and Huang 1992; Nobel 1994b). Rain roots of *A. deserti* may be 5 mm in length 5 h after rewetting, but these roots do not contribute substantially to water uptake for several days (Rundel and Nobel 1991).

Roots of desert plants are expected to have anatomical adaptations that resist desiccation and heat damage, but only a few species have been experimentally tested, notably again *A. deserti* and *F. acanthodes*. Because succulent shoot water potentials are very high, often −0.5 MPa, while soil water potential is −10 MPa or lower, one expects water to diffuse into the soil, but in succulents, water efflux is prevented because their roots act as rectifers, in a wet phase permitting rapid water uptake but in a drying soil minimizing water loss (Nobel and Sanderson 1984; Gibson and Nobel 1986; Nobel 1991c). As soil dries, roots shrink as much as 20% of original diameter, due mainly to partial collapse of cortical parenchyma, creating a soil-air gap and little direct contact between root surface and soil particles (Nobel and Cui 1992; Nobel 1994a). This soil-air gap introduces a vapor phase in the diffusion pathway that greatly slows water efflux to dry soil. Upon wetting soil, rehydration causes reexpansion of cortical cells, and may replace water in soil-air gaps, permitting rapid uptake of water again through the root surface. Resumption of absorptance by established roots is the principal means of initial water uptake for succulents 24–36 h after a rainfall (Rundel and Nobel 1991).

In *A. deserti*, root hydraulic conductivity coefficient (m s^{-1} MPa^{-1}) dropped rapidly and drastically in young roots as soils began to dry, becoming 10% of wet value, but then returned to 96% conductivity by 7 days after wetting; the decrease in water uptake was attributed in part to anatomical changes, including formation of cortical lacunae (Fig. 6.3) due to cellular dehydration and collapse, suberization or lignification of exodermis and inner cortex, and acropetal extension of suberization in endodermis (North and Nobel 1991). Older roots showed virtually no physiological or structural changes in response to drought, and no significant effects on rewetting, probably because they had already developed drought adaptations, especially suberization, that helped to limit efflux to dry soil; but roots with fully mature metaxylem experienced air embolism in tracheary elements, which also reduced potential water loss to soil. In *F. acanthodes*, 1-month-old roots showed little decrease in hydraulic conductivity but 90% recovery at 7 days after wetting, whereas conductivity was lowered one half in 12-month-old roots and recovered only 60% after wetting (North and Nobel 1992). Cortex of *Ferocactus* did not develop lacunae, but periderm was present internal to cortex and became thicker during drought. High root absorption returned in the cactus because following rehydration air embolism in vessels was substantially eliminated, and new conduits formed even during desiccation (Nobel and Huang 1992). Thus, structural modifications of root tissues, xylem embolism, soil-air gap and tissue vapor gaps, and reduced soil hydraulic conductivity play a combined role in water conservation (Lopez and Nobel 1991).

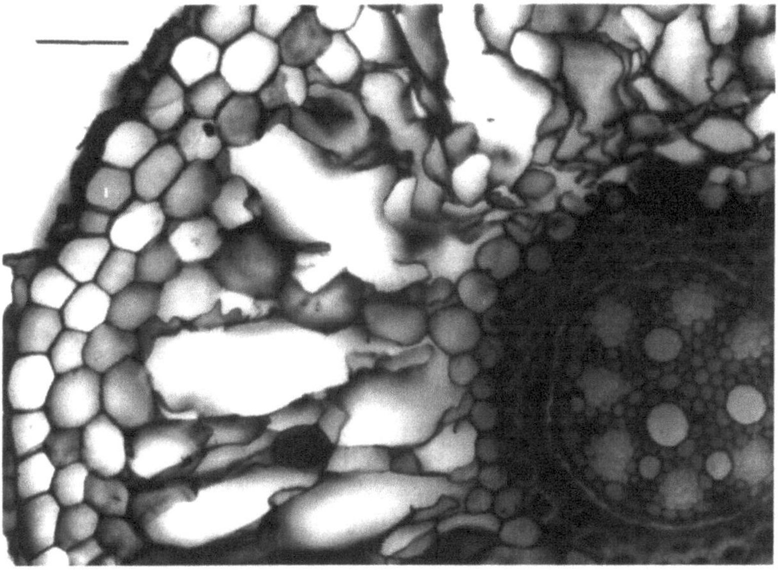

Fig. 6.3. In a 2 mm root of *Agave deserti*, cortical lacunae formed from root dehydration during drought. *Bar* 50 µm

A rhizosheath, consisting of soil particles cemented to root hairs by secreted root mucopolysaccharides (Fig. 6.4), may form to 2 mm thick on very young roots that experience a rapidly drying soil. Rhizosheaths characterize desert grasses such as *Achnatherum hymenoides* (Wullstein and Pratt 1981), species of *Stipagrostis* (Price 1911), and *Panicum turgidum* (Thomas 1921) as well as *Ferocactus*, where presence of the rhizosheath seemed to reduce potential water loss to soil, probably by forming a soil-air gap (North and Nobel 1992). Others have proposed that the rhizosheath may protect exposed roots on desert dunes from lethal temperatures and mechanical effect of sand-blast (Price 1911; Thomas 1921).

Contractile roots have been described especially from geophytes and annuals, by anatomical shortening of root axes especially during severe water stress, causing perennating organs to be pulled downward into cooler soil (Galil 1958; Zamski et al. 1983; Fahn and Cutler 1992). Nonetheless, geophytes in many deserts can survive in very shallow soil. *Lachenalia*, which is covered by a thin, sclerified leaf, is tolerant of intense heat in very poor, shallow soils, and bulbils of *Oxalis* often have hard or resin coatings, presumably an adaptation to reduce cuticular transpiration. Bulbils of *Poa sinaica* tolerate soil temperature of 80 °C and then exhibit very rapid root development after wetting (Evenari et al. 1971).

Clonal perennials may form horizontal rhizomes, connecting vascular tissues of old and young plants (Tissue and Nobel 1988). Examples in deserts are many species of *Agave*, *Ephedra*, and bunchgrasses. Clonal populations of *Larrea tridentata* in California are perhaps the oldest living plants (Vasek 1980).

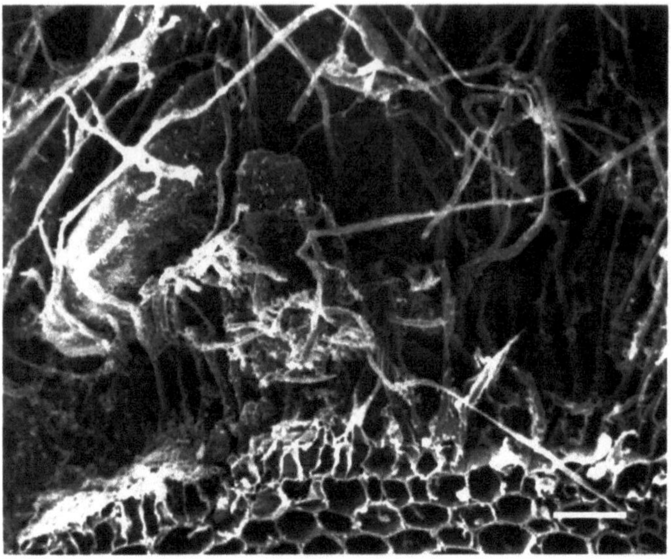

Fig. 6.4. *Achnatherum hymenoides* rhizosheath, in which soil particles are bound to root hairs by secreted mucopolysaccharides. *Bar* 50 μm

6.2.2 Leaves and Stems

In general, water enters plants through roots, flows through xylary conduits to organs along a water potential gradient, and evaporates from aerial parts having higher water vapor concentration than surrounding air, but there are climatic conditions, notably fog and dew in zones with dry soils, whereby the driving force becomes reversed, so that foliar water uptake would be theoretically possible. Foliar water uptake for desert plants has long been discussed (e.g., Volkens 1887; Holm 1894), but evidence demonstrating this is limited to few taxa, and typically is measured simply by uptake of tritiated water or dusk versus dawn differences in weight (von Willert et al. 1992). Bona fide examples should take up enough atmospheric water to have significant physiological effects when redistributed within the plant, and possess a structural design for gaining atmospheric water without losing equal or greater amounts when climatic conditions change (Rundel 1982). Significant amounts of water uptake would not be expected through cuticle-sealed epidermis with closed stomata, because epidermal walls that permit water influx also would be prone to efflux. Presence of leaf hydathodes in *Crassula* appear to be an anatomical adaptation for absorbing atmospheric water from dew (Fahn 1979).

For the Atacaman phreatophyte *Prosopis tamarugo*, laboratory evidence showed that epicuticular wax on seedling leaflets entrapped dew, fitting a popular hypothesis favoring atmospheric uptake by leaflets (Serrato Valenti et al. 1989). This study proposed that hydrophilic polysaccharides in mesophyll id-

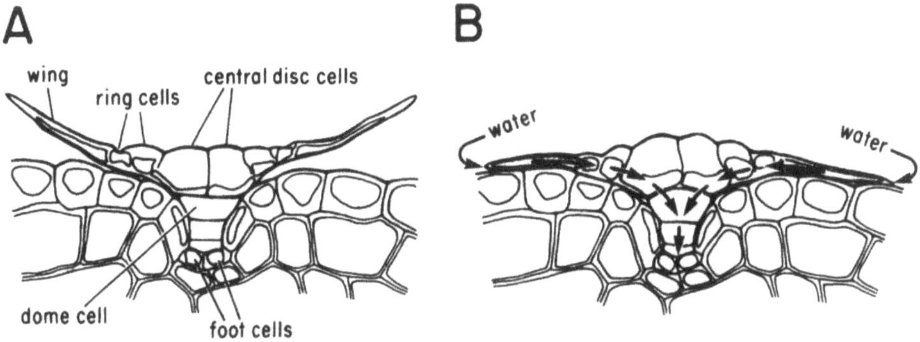

Fig. 6.5. Water-absorbing peltate trichome of *Tillandsia*. **A** Dry leaf surface, on which wings of the trichome are raised. **B** Wetted leaf surface causes water to form a film between the trichome shield and epidermis. Absorptive trichomes operate like a one-way valve; dead shield cells readily absorb moisture, and as they fill their cell walls swell to cause the shield to flatten against the leaf. Water then follows a diffusion pathway (*arrows*) along the cell lumens and across uncutinized walls through the stalk cells and into the leaf. With drying, cells no longer store water, and the outer ring cells flex upward, pinching shut cell lumens and effectively preventing water loss by the same route. (After Benzing 1970, 1990)

ioblasts and epidermal walls are involved in water uptake from the atmosphere. However, experiments have not been performed to demonstrate physiologically significant changes in chlorenchyma water potential, stem heat loads, transpiration, or biomechanical properties attributable to these walls, given that bulk water is provided by deep roots of established plants (Mooney et al. 1980b).

6.2.2.1 Bromeliaceae

Premiere examples of foliar water uptake are New World Bromeliaceae, especially rootless epiphytic and terrestrial species of *Tillandsia* in fog deserts. As reviewed recently (Benzing 1980, 1990; Rundel 1982; Fahn and Cutler 1992), covering the leaf are broad peltate trichomes, which are flexed upward under dry conditions, when wetted absorb water into dead cells, and flatten on the water film, drawing water into living stalk cells (Fig. 6.5). When surface water evaporates, the shield dries and bends upward again, sealing the stalk by a mechanism that works like an efficient one-way valve. Stomata are hidden beneath abaxial trichomes, which create a complex boundary layer and filter incoming solar radiation. Some of these species can take up water vapor from an unsaturated atmosphere (De Santo et al. 1976).

Most aridland species of *Tillandsia* appear to be CAM plants (Martin and Siedow 1981; Martin and Peters 1984) and have mesophyll consisting of isodiametric cells with substantially thickened cell walls.

6.2.2.2 *Selaginella lepidophylla* and Other Poikilohydric Plants

In deserts, water uptake has been clearly documented for microphylls of *Selaginella lepidophylla*, desert spikemoss of North America, a poikilohydric pteridophyte that exhibits photosynthetic activity within 3 h of hydration and net CO_2 within 8 h (Eickmeier 1979, 1980). By that time, microphylls are fully illuminated, and upper mesophyll has changed from spherical, desiccated cells to multilayered palisade parenchyma; by 10 h, chloroplasts already show starch accumulation (Bergstrom et al. 1982).

Selaginella lepidophylla and a few other desert species in North America are remarkable as one of a very small number of vascular cryptogams having vessels with simple perforation plates (Duerden 1934), replacing tracheids in its stems (Fig. 6.6). This has been interpreted as an adaptation to facilitate water uptake of the resurrection plants (Carlquist 1975).

Foliar water uptake, but slower photosynthetic responses, characterized related species as well as desert ferns that live in more protected sites, e.g., species of *Notholaena* and *Cheilanthes* (Harten and Eickmeier 1987), *Ceterach officinarum* (Rouschel 1938), and *Pteris*. These xeric ferns occur under rock outcrops and other special microhabitats where insolation is low and some soil moisture may be trapped. In *N. parryi* of the Sonoran Desert, fronds were adapted to shade and 90% light-saturated at 0.05% full sun, and photosynthetic temperature optimum ranged from 13–19 °C during the year, i.e., cool season or early daytime conditions (Nobel 1978a). Such pteridophytes have thin, essentially hypostomatic, weakly bifacial leaves with idioblasts in the adaxial epi-

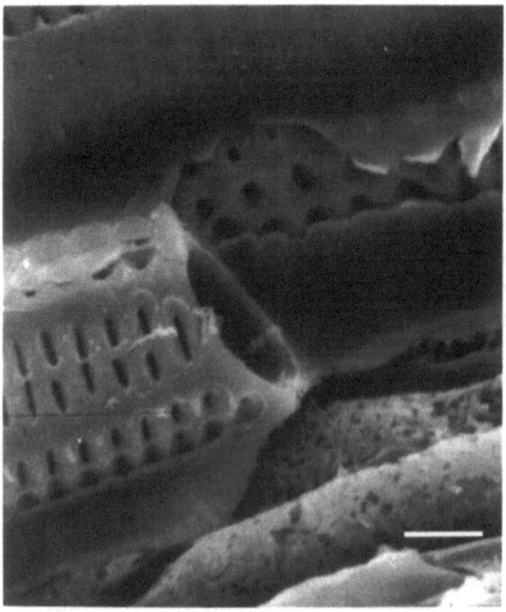

Fig. 6.6. *Selaginella bigelovii*, one of several desert species having vessels with simple perforation plates in stem xylem. *Bar* 10 μm

dermis that somehow play a role in leaf segment rolling, when dried below leaf water potential of − 1.5 MPa (Knoblock and Volz 1964; Nobel 1978a).

6.2.2.3 Water-Absorbing Stem Epidermis

Certain types of exposed cell walls on shoots are alleged to be structurally and chemically designed as rectifiers to absorb water through cuticularized cell walls, thereby improving shoot water status, but then releasing water slowly to reduce cuticular transpiration during drying (Lyshede 1977b,c; Danin 1983). Cited examples are special periclinal walls of stem epidermis of photosynthetic stems in Chenopodiaceae, Fabaceae, Brassicaceae, and Polygonaceae, and trichomes of Fabaceae with pectin-rich walls.

6.2.3 Parasitic Vascular Plants

6.2.3.1 Mistletoes

Mistletoes (Viscaceae and Loranthaceae) are infrequently observed in lowland desert-scrub habitats because their host plants typically are trees or tall shrubs (Harris et al. 1930), which tend to occur only where roots can tap deep water. Rarely, they infect tall stem succulents (Mauseth et al. 1984; Mauseth 1990). Both host types would be reliable sources for water year-round, but have low concentrations of nutrients.

At least foliose parasites, such as species of *Phoradendron* in the Sonoran Desert, are alleged to receive no host carbohydrates (Hull and Leonard 1964), but substantial heterotrophic carbon gain has been observed even without phloem connections (Marshall and Ehleringer 1990). Epiparasites are xylem-tapping because they require water and minerals, especially nitrogen. High transpiration rates during the day, from amphistomatic leaves or stems with stomata and cracks through epidermis (Calvin 1970), result in a steep water potential gradient favorable to hydrating the mistletoe (Ullmann et al. 1985), and mistletoes also have higher osmotic pressures than their hosts. Even during the night, parasite conductances are greater, probably because water vapor would diffuse from new cracks in stems and nonfunctional stomata that become necrotic in the open position (Calvin 1970). Direct xylem-to-xylem continuity between host and parasite has not been adequately demonstrated in any desert mistletoe association, and an apoplastic nontracheary pathway may be operating (Stewart and Press 1990).

Nitrogen is supplied to the parasite in the xylem stream; mistletoes growing on nitrogen-fixing hosts contain high levels of nitrogen in their tissues, and their high transpiration rates, hence high water demands, appear instead to represent a nitrogen-gathering mechanism (Schulze and Ehleringer 1984; Ehleringer et al. 1985, 1986).

No physiological studies have been conducted on the endophyte of *Tristerix aphyllus*, but this species had only a few filaments within xylem of its succulent cactus host, but an abundance of parasitic tissues within phloem (Mauseth 1990). Likewise, the endophyte of the diminutive stem parasite *Pilostyles thurberi*, which has sunken abaxial stomata on perianths, is primarily developed within the secondary phloem of the host, and nonvascularized filaments (sinkers) extend through host xylem (Rutherford 1970). If stomata are functional, they would be adaptive for increasing nutrient flow for emergent reproductive structure.

6.2.3.2 Root Parasites

Thick, fleshy root parasites of North American Lennoaceae lack any adaptations to restrict water loss from achlorophyllous stems and leaves. In *Pholisma arenarium* and *P. sonorae*, epidermis on a shoot mostly lacks cuticle, thus making this a major pathway for transpiration to aerial portions and belowground stems in contact with dry soil. Very wide vascular bundles supply the flowering shoot with water via more than 2000 extremely narrow vessels (rarely exceeding 10 μm lumen diameter) formed by fascicular vascular cambium. Water for the infected host-plant roots probably is drawn from a wet soil stratum, e.g., in deep sand, but ecophysiology of these species has never been studied. Likewise, plant water relations of achlorophyllous desert Orobanchaceae are unknown.

6.3 Vascular Tissues

One of the challenges of desert botany is to elucidate interrelationships of leaf physiology and shoot phenology with xylem structure and to maintain functional water conduits while avoiding damage caused by heat and wind.

6.3.1 Veins in Leaves

Some generalizations about vascular tissues in microphylls of nonsucculent desert dicotyledons can be made: (1) transectional area of xylem and phloem is small relative to plants from other habitats with even thinner leaves; (2) vessels are extremely narrow, and there are few vessels per bundle; (3) sieve tubes are extremely narrow; (4) virtually no secondary growth has been observed in vascular bundles of most veins; and (5) vascular bundles tend to be close, maintaining a relatively short distance for water to diffuse from vessels to mesophyll or sucrose to be transported from mesophyll to sieve tubes. Broad leaves, including winter- and summer-active herbs, phreatophytes, and palms, have much more conducting xylem to supply greater leaf area.

Limited development or absence of secondary growth in lamina, along with absence in many of bundle sheath extensions of sclerenchyma or collenchyma, explains why veins of many desert microphylls are not conspicuous on the surface. Exceptions with prominent surface veins are most often observed in Asterales, Lamiales, and Malviflorae. Without abaxial and adaxial bundle sheath extensions, a leaf can have much higher A^{mes}/A, hence higher CO_2 fixation, perhaps even greater PPFD absorptance, and without extra mechanical tissues, structural costs per leaf are reduced (Orians and Solbrig 1977a).

Typical metaxylem vessels of dicotyledonous leaves have secondary helical thickenings. Some of the species experiencing extremely low leaf water potentials (below -6.0 MPa) have metaxylem elements with pitted lateral walls, e.g., summer-active woody plants, such as evergreens, C_4 species, and *Artemisia herba-alba*. A very narrow, pitted metaxylem vessel may help to prevent cavitation.

Many angiosperms have imperforate tracheids, terminal tracheids with lignified secondary cell-wall thickenings, located at vein endings (areoles) of leaves. In plants from dry habitats, clusters of tracheoidioblasts have been described at vein endings, especially in halophytes (Gamaley 1984) and penetrating stem chlorenchyma (Fahn and Arzee 1959). In *Tetradymia*, which exhibits stem photosynthesis, tracheoids lie adjacent to bundle sheath of the stem (Böcher 1971). Traditionally, authors have suggested that these cells provide a water-storage function, and have assumed that these help to rehydrate neighboring chlorenchyma (e.g., Volkens 1887), and for tracheoidioblasts with protoplasts may play a role in water relations in halophytes (Weber et al. 1977) or as part of a water-uptake system from dew (SaadEddin and Doddema 1986). To date, the water-storage function of empty tracheoidioblasts has not been proven, and no explanation accounts for the presence of secondary thickenings in dead cells versus thin walls of living water-storage parenchyma. In species of *Sansevieria* from semiarid habitats, imperforate cells reinforced with secondary thickenings provided anatomical hinges for dehydrated leaves to avoid permanent physical damage to adjacent living tissues during collapse (Koller and Rost 1988a,b).

No study has pinpointed special desert patterns of leaf vasculature, although smaller intervenal distances reduce the diffusion distance from primary xylem to leaf parenchyma (Shields 1950). Mean intracostal distances of desert microphylls tend to be 75–150 µm.

6.3.2 Secondary Xylem

Customary methods of comparative wood anatomy often employ standardized sampling by describing outermost sapwood of the lowermost stem. Such studies have identified the following wood features of the majority of desert shrubs (Webber 1936; Carlquist 1975, 1984, 1985; Carlquist and Hoekman 1985; Baas and Carlquist 1985; Fahn et al. 1986): (1) growth rings distinct; (2) vessel density high per transectional area; (3) vessels relatively narrow to extremely narrow; (4) vessel elements exclusively with simple perforations plates (foraminate in *Ephedra*); and (5) vessel elements short, sometimes extremely short. From a phylogenetic viewpoint, these apomorphic features clearly indicate that desert taxa are members of highly specialized lineages (Carlquist 1975). None of the families retaining primitive dicotyledonous features has been successful in colonizing warm deserts. Consistently, wood anatomists have recognized phreatophytes, especially trees growing along wadis, as exceptions, because they characteristically have wider vessel lumens and much lower vessel densities.

For secondary xylem hydraulic conductance (k_h) for unobstructed vessels with either simple or compound perforation plates can be accurately estimated using the Hagen-Poiseuille equation:

$$k_h = \frac{\pi \sum^{n} d_i^4}{128\,\eta} \qquad (6.1)$$

where η is vicosity of xylem sap and d_i is vessel lumen diameter (m), modified for elliptical outline (Gibson et al. 1984). Predicted by this equation, comparing a vessel with a lumen diameter of 40 µm to one with 20 µm diameter, water flow is 16 times greater (d_i^4) in the wider one under the same water potential gradient; therefore, under nonstressed conditions, relatively few unobstructed wide vessels will transport nearly all of the water for an axis.

Carlquist and Hoekman (1985) have interpreted the typical desert wood as having arisen under selection for safety during periods of water stress while sacrificing high xylem conductance (Zimmermann 1983; Baas et al. 1983; Carlquist 1984). In this model, wide vessels are assumed to be more prone to cavitation by developing air emboli, so that abundant narrow vessels are an alternative strategy to enable water conductance to continue even when many of the vessels have become dysfunction.

Carlquist and Hoekman (1985) concluded from a study of Californian native plants that three additional features are characteristic of desert shrubs: vessel grouping tends to be high; vessel groups often contain specialized tracheids (fiber-tracheids, short tracheids, vasicentric tracheids, or vascular tracheids); and vessels frequently have tertiary helical thickenings on lateral walls (40% of the warm desert examples in that study). These vascular features were interpreted as a basic xylem strategy to reduce vulnerability of plants of dry habitats to water stress:

1. Many narrow vessels confer greater safety than wider ones because water flow is less affected if some of the vessels become dysfunctional via embolism.
2. Short vessel elements may localize air bubbles and thereby should restrict the longitudinal extension of embolism.
3. Imperforate tracheary elements having bordered pits, when associated with a vessel group, can provide an alternative water pathway around air emboli in vessels.

4. Helical sculpture may help to maintain the integrity of vessel flow during drought.

Desert woods surveyed from the Negev showed some similar anatomical trends, but enough differences to raise concern about drawing functional conclusions without substantial physiological tests (Baas and Carlquist 1985; Fahn et al. 1986). There, for example, vessels with tertiary helical thickenings on lateral walls mostly grow in rock crevices, not in valley habitats. From both deserts, examples with tertiary vessel sculpturing largely occur in particular families (Asteraceae, Fabaceae/Faboideae, Lamiaceae, Rhamnaceae, Rosaceae, and Solanaceae), suggesting a phylogenetic rather than a functional explanation may be a more robust explanation, i.e., tertiary helical thickenings arose in other semiarid habitats, where origin of sculpturing first needs to be understood. Nonetheless, tertiary helical thickenings appear in most woods of green-stemmed plants, these from numerous plant families (Table 4.1).

Species in many different families have converged to a condition wherein vessels are in contact with relatively thin-walled imperforate tracheids. Carlquist (1980) has reasoned that imperforate conducting cells would be less vulnerable to air emboli and thereby provide a pathway within each vessel group around air-blocked vessels. This hypothesis has not been tested with water-flow and embolism studies for vessel-bearing woods from deserts.

Thick-walled vessels have been reported for plants that experience very low shoot water potentials during drought, and the feature has been interpreted as an adaptation to limit collapse of vessels when placed under great tension (Carlquist 1975). This hypothesis would not appear to be correct, simply given that leaves, which experience the most negative water potentials, have primary xylem vessels with walls that are not excessively thickened. Instead, vessels in woods with greatly thickened walls (>4 μm for narrow vessels and thicker for wider ones) appear to lack tyloses, whereas vessels with thinner walls may be plugged by tyloses. Perhaps thicker walls of vessels in xerophytes are an adaptation to stop tylosis formation.

The majority of desert woods in the Negev have relatively short, nucleate libriform wood fibers, and many are septate (Fahn and Leshem 1963; Fahn et al. 1986). Some species have living wood fibers that store starch. Adaptive significance of living fibers has never been explored, although authors have speculated that they play a role in reducing vessel cavitation. Living fibers have a higher degree of flexibility without breaking and may be specialized to increase bending for withstanding strong desert winds.

6.3.2.1 Shrubs and Subshrubs with Drought-Deciduous Leaves

Typical mean vessel diameters of woody plants with drought-deciduous leaves tend to be <40 μm at densities of 180–450 mm^{-2} (Carlquist and Hoekman 1985; some of the densities published in this article appear to be too low). Narrow vessels that characterize first-formed secondary xylem of young shoots are strongly related to narrow vessels entering typical drought-deciduous microphyllous leaves, some of which are able to transport enough water to transpiring leaves at very low leaf water potentials, e.g., −7.5 MPa in *Artemisia herba-alba* (Kappen et al. 1972). Species with broader leaves tend to have fewer, wider vessels, e.g., *Encelia farinosa* (40.3 μm, 108 mm^{-2}), but this is not always the case.

Among drought-deciduous phanerophytes and chamaephytes are taxa with long shoot-short shoot organization, where a canopy remains mostly intact during drought, versus shoot-shedders, which experience progressive basipetal desiccation and would be predicted to experience greater embolism. Nonetheless, woods of both strategies from lowland, southern Mojave desert scrub share the same range of anatomical characteristics. In the Mojave Desert, a majority of common woody taxa are Asteraceae, which have vessel grouping ranging from low to extremely high. A significant number of solitary vessels occur in woods of *Encelia farinosa*, which has seasonally dimorphic and relatively broad leaves (Chap. 2) and retains foliage during much of the year. *Lycium andersonii*, which forms relatively wide vessels in tangentially arranged clusters, has thick fleshy leaves.

Most xerohalophytic woody plants are deciduous species of Chenopodiaceae and have some of the shortest and narrowest vessel elements of the desert environment (Fahn et al. 1986).

6.3.2.2 Evergreen Shrubs

A distinctive wood design appears in evergreen phanerophytes, e.g., *Larrea tridentata, L. divaricata, Mortonia* spp, and *Zygophyllum dumosum*, which are highly successful in their respective deserts, and *Simmondsia chinensis*, which occurs along the desert margin (Bailey 1980). In all of these, vessels are solitary or nearly so (mean vessel grouping 1.00 to 1.10; Carlquist and Hoekman 1985; Fahn et al. 1986), and their woods are diffuse porous and lack tyloses.

Larrea tridentata is used as an example having net carbon gain in all months of the year, even under midday leaf water potentials between −6.0 and −8.0 MPa (Odening et al. 1974), and which can respond to heavy summer rain within 2 weeks with vigorous new growth without accompanying growth of new wood. This species experiences substantial rehydration of shoots during nighttime, an increase of > 2 MPa. Creosotebush has a low vessel density of 100–110 mm^{-2}, all vessels narrow to extremely narrow and thick-walled, whereas imperforate xylary elements are extremely thick-walled and often filled with starch grains, hence these are unlikely conduits for water transport. Its solitary vessels may be a valuable mechanical design to reduce cavitation by stem flex during strong winds.

Desert woods with nearly all solitary vessels are anticipated in families where true tracheids or fiber-tracheids occur, as in Zygophyllaceae, Rosaceae, Celastraceae, and Simmondsiaceae (Corlquist 1982). Drought-tolerant

Table 6.1. Warm-desert species of Israel (I, from Fahn et al. 1986) and North America (NAm) having widest vessels with tangential diameters > 150 μm in lower stem sapwood. These are species with broad leaves or photosynthetic stems, or which are phreatophytes

Taxon	Phreatophytic	Broad-leaved	Photosynthetic stems
Acacia spp. I,NAm	+		
Balanites aegyptiaca I	+	+	+
Calligonum comosum I			+
Calotropis procera I	+	+	
Cercidium spp. NAm	+		+
Chilopsis linearis NAm	+		
Citrullus colycinthus I		+	
Cocculus pendulus I		+	
Cucurbita digitata NAm		+	
C. palmata NAm		+	
Datura wrightii NAm		+	
Ficus pseudo-sycomorus I		+	
Koeberlinia spinosa NAm			+
Moringa peregrina I	+		
Ochradenus baccatus I			+
Olneya tesota NAm	+	+	
Prosopis spp. I,NAm	+		
Tamarix spp. I	+		
Ziziphus spina-christi I	+	+	

evergreens do not appear to possess a backup system of tracheids for water flow during water stress, given that thick-walled tracheids are not designed as water conduits.

6.3.2.3 Phreatophytes

Although interpretations of desert woods to date have focused on safety, i.e., a negative impact on plant water relations, the obvious, but still unmeasured, comparison should be a positive one, i.e., how hydraulic conductance of a stem segment relates to area of photosynthetic canopy that it supplies, e.g., leaf-specific conductivity or Huber value (Zimmermann 1983). Plants with wide vessels in lower branches and trunks are those supplying leafy canopies, especially phreatophytic trees and broad-leaved species, or plants with green stems that possess large transpiring surface areas (Table 6.1). Phreatophytic trees have high water requirements to supply high transpiration rates; in *Acacia raddiana, A. tortilis,* and *Tamarix aphylla*, wood of trunk often contains vessels greater than 200 μm, solitary or in small multiples (Fahn 1958, 1959).

Comparing different growth habits from California, warm desert washes (Table 6.2) reveals that trees typically have relatively few, wide vessels, whereas shrubs tend to have numerous narrow vessels. Among trees, the exception is the aphyllous, green-stemmed *Psorothamnus spinosus*, which has woods similar to shrubs; among shrubs, exceptions are *Petalonyx linearis*, which is arido-

Table 6.2. Vessel characteristics of North American shrubs and trees occurring along desert washes: mean density of vessels (n mm^{-2}), mean vessel diameter (VD, μm), mean vessel-element length (VE, μm), vessels per group (V/G). (Data from Carlquist and Hoekman 1985)

	n mm^{-2}	VD	VE	V/G
Shrubs				
Asclepias albicans	196	20.9	224	2.9
Brickellia multiflora	127	55.6	163	1.5
Calliandra eriophylla	187	26.1	137	3.0
Chrysothamnus paniculatus	302	25.5	126	16.6
Fallugia paradoxa	444	20.6	208	2.0
Hymenoclea salsola	191	27.6	106	1.7
Hyptis emoryi	48.8	63.8	254	1.6
Isomeris arborea	320	21.5	111	5.1
Justicia californica	146	49.8	245	2.0
Lepidospartum squamatum	203	49.9	164	17.0
Lycium brevipes	327	16.5	226	7.2
Petalonyx linearis	36.0	63.0	258	1.2
Prunus fremontii	143	27.7	329	1.6
Salazaria mexicana	268	24.5	142	2.0
Salvia dorrii	697	19.3	162	3.2
Senna armata	182	24.6	113	1.5
Stanleya pinnata	214	26.5	153	4.4
Trees				
Acacia greggii	28.4	42.2	181	2.3
Cercidium floridum	35.6	54.2	205	2.4
Chilopsis linearis	30.1	86.7	132	1.5
Olneya tesota	25.9	80.0	183	2.9
Prosopis pubescens	68.1	47.4	162	2.9
Psorothamnus spinosus	90.4	35.9	159	7.1

active and must tap deep moisture, and *Hyptis emoryi*, a relatively broad-leaved species also with deep roots. *Petalonyx thurberi*, which has small triangular leaves and photosynthetic stems, has diffuse-porous xylem with low vessel density, having several first-year vessels >80 μm in tangential diameter.

For phreatophytes, timing of cambial activity varies among the different taxa, and amount and nature of secondary xylem changes accordingly. In *Acacia raddiana*, *A. tortilis*, and *Tamarix aphylla*, cambial activity occurs throughout the year, wood is diffuse-porous, and the growth rings are indistinct (Fahn 1958, 1959), but cambial activity is poorly correlated with leaf growth (Halevy and Orshan 1973) and each species responds in different ways to water, high temperature, and photoperiod (Fahn and Sarnat 1963; Fahn et al. 1968). Cambial activity of *Ziziphus spina-christi* growing under dry conditions, instead, is limited to relatively short periods when young leaves and flower buds formed in April and May, but extended for half a year when roots tap the water table and produced two to three times more secondary xylem (Liphschitz and Waisel 1970a). In *Populus euphatica*, plants of dry habitats had ring-porous wood whereas those of wet habitats had diffuse-porous wood (Liphschitz and Waisel 1970b). In the Sonoran Desert, radial growth of trees was correlated with habitat (Turner 1963).

6.3.2.4 Nonsucculent Aphyllous Shrubs and Trees

Relatively high calculated values of hydraulic conductance are obtained from aphyllous green-stemmed species as compared with drought-deciduous species having microphylls. This reveals that aphylly has been incorrectly interpreted as an adaptation to reduce transpiration to extremely low levels; quite the contrary, green-stemmed species draw water from deep layers to maintain photosynthetic activities of stems. Some woods of green-stemmed phanerophytes have a high proportion of solitary vessels and may be ring-porous (Gibson 1979).

In *Retama raetam*, horizontal surface roots have much wider vessels, hence greater hydraulic conductance, than vertical deep roots (Fahn 1964). Such dimorphic roots have been explained as a mechanism to harvest rainwater that briefly infiltrates shallow soils. In other green-stemmed species, widest vessels in the plant are found in the large main root. For example, very large specimens of *Castela polyandra* and *Koeberlinia spinosa* have ring-porous trunk wood, but its diffuse-porous main root wood has wider vessels up to 100 μm in diameter, sufficient, one assumes, to hydrate the stem surface of a large canopy.

Flexible green stems of the gymnosperm *Ephedra* have solitary vessels but with foraminate perforation plates and thick-walled tracheids (Carlquist 1988; Fahn et al. 1986). Such tracheids do not appear to be designed for physiologically significant amounts of water transport.

6.3.2.5 Succulents

Succulent species year-round maintain high stem water potentials (Chap. 5) and tend to have secondary xylem that is anatomically very different from that of nonsucculent desert shrubs and trees. Because transpiration rate is extremely low, for either CAM or C_3 species, diurnal fluctuations in stem water potential are extremely small; in succulents, xylem tensions most likely are low, hydraulic conductance routinely will be low, and danger of cavitation probably is not a common problem. Greatest change in stem water potential occurs when a droughted plant receives a sudden rainfall, followed by rapid water uptake that restores stem water within a 24-h period (Gibson and Nobel 1986).

Vessels in stem succulents are unusual in that they tend to have elongate pitting on lateral walls (Fig. 6.7A), and vessel elements tend to be much longer and wider and have thinner walls than in other desert woods. Examples would include columnar cacti and shrubby platyopuntias of American deserts and shrubby succulent euphorbs of Old World deserts. Presence of elongate pitting and longer elements was used to explain these occurrences as paedomorpho-

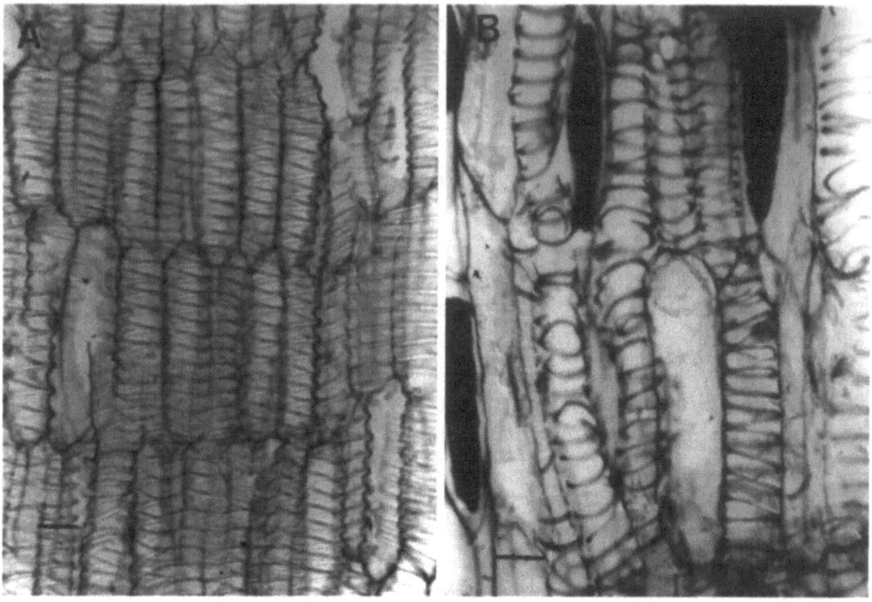

Fig. 6.7A,B. Characteristic lateral walls of succulent woods (secondary xylem) in low growth habits. **A** *Echinocactus polycephalus*; radial view of imperforate vascular tracheids with secondary helical thickenings. **B** *Dudleya* aff. *saxosa*; radial view of vessels with secondary helical thickenings. Magnification bar = 10 μm

sis (Carlquist 1962, 1975; Gibson 1973). Presence of large pit membranes, as on vessels with scalariform pitting, would be expected to facilitate horizontal diffusion but has not been tested.

Shrubby cylindropuntias on North America (*Opuntia*) have vessel elements with small, multiseriate alternate pitting embedded in a hard matrix of nucleate libriform fibers. Tyloses sometimes are observed in some of these woods, suggesting that cylindropuntias are more prone to form emboli than other cacti that have more stem-succulent tissues.

In Cactaceae, evolution has produced very succulent stem forms having secondary xylem with large patches of unlignified cells, permitting water storage within wood and easier pathways for diffusion from vessels to succulent tissues in the pith and cortex (Gibson 1973, 1975, 1977, 1978b,c; Gibson and Nobel 1986). Barrel cacti (e.g., *Ferocactus* and *Echinocactus*), low caespitose forms (e.g., *Echinocereus, Mammillaria dioica,* and *Opuntia basilaris*), and solitary, low cylindrical to globular individuals (e.g., *Coryphantha*) have nonfibrous xylem, i.e., lacking libriform wood fibers; as a replacement for wood fibers, and presumably homologous with them, occur vascular tracheids, fusiform imperforate cells with scanty helical or annual secondary thickenings (Fig. 6.7A).

Nonfibrous stem woods in cacti are mechanically weak but can be tolerated in small growth forms that rely on support from succulent tissues of pith and cortex (Gibson and Nobel 1986). With little secondary cell-wall content, the xylem is energetically less expensive to make. Vascular tracheids, which can store water, also provide little resistance to diffuse in any direction. Most importantly, flexible secondary thickenings in vascular tracheids can expand and contract with changes in water availability without damage to cells. Roots of these species typically have nucleate, septate libriform fibers throughout secondary xylem.

No other family of stem succulents has evolved nonfibrous woods with vascular tracheids like those of Cactaceae, but lateral-wall pitting has been replaced by helical or reticulate secondary thickenings in woods of certain Crassulaceae (Fig. 6.7B; Carlquist 1975), Aizoaceae (Gibson 1994), and several other lineages.

Development of succulence and nature of woods in stems and roots of Fouquieriaceae have been carefully described and discussed from a functional viewpoint (Henrickson 1969a,b,c, 1972, 1975, 1977). Whereas *Fouquieria columnaris* is a true stem succulent, with massive water storage tissue and woods with paedomorphic features, those of *F. splendens* and other ocotillo species have none of the characteristic anatomical wood features of stem succulents. Coupling observations that ocotillo produces new mature leaves from short shoots within a few days of an effective rain yet abscises leaves fairly rapidly as a plant enters drought, a logical conclusion is that the plants have high conductivity but low safety.

In large stems of the pachycaulous arborescent succulents *Pachycormus discolor* (Gibson 1981) and *Bursera microphylla* (Carlquist and Hoekman 1985), vessel densities are low and vessels are wide, similar to those in nonsucculent phreatophytic trees, equipped for high xylem conductances to rehydrate a substantial canopy of microphyllous leaflets. Relatively thin fiber walls make these succulent woods very soft, and this probably contributes substantially to horizontal water diffusion within the plant axis.

6.3.2.6 Herbaceous Angiosperms

Herbaceous winter annuals tend to be diminutive plants with tiny leaves, hence small leaf area and little need for high hydraulic conductance of the xylem. Even the cool-season broad-leaved species tend to have no special vascular adaptations, but they perish when xylem formed in early spring cannot effectively rehydrate leaves with the onset of drought.

During hot summer months, extremely high xylem hydraulic conductance is required for transpirational cooling to maintain very wide leaves (Smith 1978; Chap. 2). It is not surprising, therefore, that the widest vessels of first-year secondary xylem of all desert plants are found in the broad-leaved arido-active hemicryptophytes, e.g., *Datura wrightii* (maximum tangential vessel diameter 155 μm), *Asclepias erosa* (110 μm), and a desert cucurbit *Cucurbita palmata* (195 μm). In comparison, tangential vessel diameters in first-year xylem of phreatophytes, which eventually have wide vessels in lower stems, rarely exceed 70 μm.

In a desert palm leaf, which lacks secondary growth, many vessels with diameters > 50 μm supply water to the leaf rachis.

6.3.3 Split Axis

Hemispherical desert chamaephytes, which have many basal branches, may experience splitting so that an old main root-shoot axis separates into somewhat independent segments comprising a single mature canopy (Ginsburg 1963; Fahn 1964; Evenari et al. 1971; Jones and Lord 1982; Fahn and Cutler 1992). Segments may be partially or almost entirely isolated anatomically by formation of interxylary cork (*Artemisia herba-alba* and *Achillea fragrantissima*), sectoral cessation of cambial activity (*Zygophyllum dumosum*), or formation of basipetal splits from branch axils to between flutes of the vascular cylinder associated with diverging branch (*Ambrosia dumosa, Peganum harmala*, and *Acamptopappus sphaerocephalus*; Fig. 6.8). For old specimens of *Z. dumosum*, discontinuities in cambial activity have been correlated to drought events (Danin 1983).

Theoretically, when physically isolated branches should have more physiological independence, so that during drought a plant could experience the death of one or more segments while the remainder with more favorable water status, i.e., rooting in the most favorable microclimate, could survive (Ginzburg 1963; Fahn 1964). To date, the only experiments performed found no consistent differences in stem water potentials followed monthly for individual

Fig. 6.8. *Acamptopappus sphaerocephalus*; a basal, old stem exhibiting a well-developed split axis

branches of *A. dumosa* (Jones 1984), but a long-term study would be required to follow death of tagged branches to determine whether shoots that die are more prone to water stress. Effects of wind on axis splitting also need to be examined using biomechanical methods.

6.3.4 Anomalous Secondary Thickening

Several reviews of desert plant adaptations (Fahn 1964; Fahn et al. 1986; Fahn and Cutler 1992) have paid special attention to anomalous secondary thickening, in particular because each desert has several common representative species. Chief among these are Chenopodiaceae, Amaranthaceae, Nyctaginaceae, and Aizoaceae in the order Caryophyllales; all species of these four families have anomalous secondary thickening by successive cambia in stems and roots (Gibson 1994). Aizoaceae have succulence and CAM (Chap. 5), and many of the other families are C_4 species, especially among halophytes and summer herbs.

Certainly, Chenopodiaceae and Aizoaceae experienced early radiations in desert habitats, perhaps suggesting that anomalous growth may be adaptive, but there are no data from experiments showing desert advantages of these axes versus those with normal secondary thickening. Possible functional benefits could be any of the following.

1. The vascular system permits branch-shedding without endangering the life of the main plant axis.
2. Functional sieve tubes are positioned between arcs of water-conducting xylem, potentially benefiting sugar translocation.
3. Extremely short, often very narrow vessel elements, produced from new, fusiform initials for each successive cambium, avoid problems of air embolism in extremely dry soils.
4. Arcs of xylem produced by successive cambia appear to form a reticulum, as do arcs of sieve tubes, perhaps improving delivery of water and sugars, respectively.
5. Functional phloem is deep-seated and thereby protected from desiccation or temperature damage close to the stem surface.

In vascular tissues of desert Caryophyllales, vessel elements are extremely short, in great part due to the manner in which storied initials of successive cambia arise from short, fusifusm secondary parenchyma cells (Gibson 1994), and they are extremely narrow (Carlquist and Hoekman 1985; Fahn et al. 1986). Vessel elements form clusters because they are formed only within a short arc and only opposite sieve tubes, but it is misleading to classify secondary parenchyma or paratracheal parenchyma as being vasicentric tracheids (Carlquist and Hoekman 1985), which they are not. The arido-active evergreens *Atriplex hymenelytra* and *A. canescens* have thick vessel walls, and no tyloses form.

Woody axes of chenopods, such as *Atriplex* and *Grayia*, and ice plants, such as *Ruschia* and *Brownanthus*, can be extremely hard, due to vascular sclerenchyma, but brittle because elements are extremely short. Abscission of the entire shoot through a weak zone in vascular tissues has evolved in tumbleweeds, *Salsola* and *Kochia*, for wind-driven dispersal (Becker 1968, 1978). Secondary parenchyma within stems of succulent and fleshy stems and roots, as in Aizoaceae and *Abronia*, act as water-storage tissues.

Anomalous secondary growth is also pronounced in the North American evergreen *Simmondsia chinensis* (Bailey 1980), the Old World *Salvadora persica* (Evenari and Gutterman 1973), and many Capparaceae (Metcalfe and Chalk 1950).

6.3.5 Ground Tissues

Desert plants of the Negev tend to have relatively narrow pith in young dicotyledonous stems and a high cortex/stem ratio, suggesting that a thicker cortex affords some protection to the stele before periderm is formed (Fahn 1964; Fahn and Cutler 1992). Especially in desert Chenopodiaceae, cork tissue forms deep in cortex close to vascular tissues and has been hypothesized as an adaptation against damage to phloem or xylem (Fahn 1990).

7 Origins of Desert Structural Adaptations

Most indigenous perennial plants of warm desert ecosystems apparently evolved from ancestors in subtropical and tropical semiarid habitats, especially drought-deciduous scrub and thorn forest for succulent and nonsucculent woody dicotyledons and grasslands for C_4 bunchgrasses (Hattersley 1983; Shmida 1985; Gibson and Nobel 1986). Woody perennials tend to experience severe dieback or are killed when exposed to tissue-freezing temperatures, and having a warm-climate ancestry probably is why most have not evolved biochemical adaptations to tolerate freezing. Somewhat resistant to cold temperature stress are many woody Rosaceae, Chenopodiaceae, Asteraceae, and *Ephedra*, which tend to occur in colder montane habitats and are commonly observed also in cool deserts. Annuals, C_3 grasses, and geophytes, especially those active in cool seasons, often have closest phylogenetic affinities with lineages in semiarid warm temperate or temperate habitats, e.g., sclerophyll and drought-deciduous scrub, and steppe, where some freezing tolerance would be expected.

7.1 Where Desert Adaptations Evolved

Plants that have adapted to warm desert climates possess distinctive vegetative designs affecting energy budgets (Chaps. 2 and 4), gas diffusion rates (Chaps. 3 and 4), plant water relations (Chaps. 5 and 6), phenology, and biochemical responses to stress. Through that selective process, relatively few plant families have contributed species, so that less than 30% of seed plant families (127 of 456) are represented in warm desert floras (Table 7.1). Of these, only one-quarter are represented by more than five genera, and many of the listed families are represented by one or only a few species.

A distinction should be made between structural adaptations that originated in populations differentiating within the desert environment versus characters preexisting within lineages of semiarid climates and thereby preadapted for desert life. The best-studied case of an adaptation that arose specifically within the desert is found in *Encelia farinosa*, which forms leaves with dense covering of IR-reflective trichomes (Chap. 2). Its heteroblastic structural adaptations, without other physiological modifications, have enabled this species to succeed in the warm desert environment (Ehleringer and Mooney 1978).

Table 7.1. Vascular plant families having representatives (excluding aquatics) in warm deserts of the world, i.e., an arid habitat with virtually no freezing temperatures. Angiosperm families are those recognized by Thorne (1992), and parentheses indicate fraction of families represented in warm deserts. Families in boldface are those having at least six genera represented in scrub or sane-dune habitats of warm deserts. Many of the small families are also heavily represented in other arid and semiarid habitats

PTERIDOPHYTES

Pteridaceae (*Cetarch, Cheilanthes, Notholaena, Pellaea,* and *Pentagramma*)
Selaginellaceae

GYMNOSPERMS (4/18)

Cupressaceae (*Juniperus* marginally)
Ephedraceae
Pinaceae (*Pinus* marginally)
Welwitschiaceae

ANGIOSPERMS

Dicotyledons (103/352)
Acanthaceae
Achatocarpaceae
Aizoaceae
Amaranthaceae
Anacardiaceae
Apiaceae
Apocynaceae
Asclepiadaceae (incl. Periplocaceae)
Asteraceae
Balanitaceae
Berberidaceae
Bignoniaceae
Boraginaceae
Brassicaceae
Burseraceae
Cactaceae
Calyceraceae
Campanulaceae
Capparaceae
Caryophyllaceae
Casuarinaceae
Celastraceae
Chenopodiaceae
Cistaceae
Cochlospermaceae
Convolvulaceae
Crassulaceae
Crossosomataceae
Cucurbitaceae
Cynomoriaceae
Dipsacaceae
Ebenaceae
Euphorbiaceae
Fabaceae
Fouquieriaceae
Frankeniaceae (xerohalophytes)
Geraniaceae
Globulariaceae
Goodeniaceae
Halophytaceae
Hydrophyllaceae
Koeberliniaceae
Krameriaceae
Lamiaceae
Lennoaceae
Loasaceae
Loganiaceae
Loranthaceae
Malesherbiaceae
Malpighiaceae
Malvaceae
Martyniaceae
Meliaceae (marginally)
Menispermaceae (marginally)
Molluginaceae
Moraceae
Moringaceae
Myrtaceae
Nesogenaceae
Neuradaceae
Nolanaceae
Nyctaginaceae
Olacaceae
Oleaceae
Onagraceae
Oxalidaceae
Papaveraceae
Passifloraceae
Pedaliaceae
Plantaginaceae
Plumbaginaceae
Polemoniaceae
Polygalaceae
Polygonaceae
Portulacaceae
Proteaceae
Rafflesiaceae
Resedaceae
Rhamnaceae
Rosaceae
Rubiaceae
Rutaceae

Table 7.1 (*contd.*)

Salicaceae (marginally)	MONOCOTYLEDONS (20/86)
Salvadoraceae	
Sapindaceae	Agavaceae
Scrophulariaceae (incl. Orobanchaceae)	Alliaceae
Simaroubaceae	Alstroemeriaceae
Simmondsiaceae	Amaryllidaceae
Solanaceae	**Arecaceae** (marginally)
Stegnospermaceae (marginally)	Asparagaceae
Sterculiaceae	Asphodelaceae
Stylidiaceae	Bromeliaceae
Tamaricaceae	Calochortaceae
Thymelaeaceae	**Colchicaceae**
Tiliaceae	Cyperaceae
Tropaeolaceae	Dasypogonaceae
Turneraceae	Dioscoreaceae
Ulmaceae	Dracaenaceae (incl. Nolinaceae)
Urticaceae	**Hyacinthaceae**
Verbenaceae	Iridaceae
Viscaceae	Ixioliriaceae
Vitaceae	Liliaceae
Zygophyllaceae (incl. Nitrariaceae)	**Poaceae**
	Tecophilaeaceae

Aphyllous shrubs, which utilize stems as the chief assimilatory organ (Chap. 4), apparently also evolved within the desert environment from leafy ancestors in wetter habitats; examples in South America are *Monttea aphylla* versus the leafy *M. chilensis* and *Bulnesia retama* versus leafy species *B. chilensis* or *B. foliosa*. In North America, the genus *Castela* includes green-stemmed desert shrubs (formerly *Holacantha*), *C. emoryi*, *C. polyandra*, and *C. stewardii*, versus the leafy species in semiarid habitats, *C. tortuosa* and *C. peninsularis*. Wide-ranging species that have populations in deserts as well as semiarid habitats, such as phreatophytic trees (Halevy and Orshan 1972), generally are assumed to be fairly recent desert additions involving few or no modifications in fundamental physiological, phenological, and structural properties.

Among succulents are numerous cases for origin of desert features via preadaptation, and structural adaptations were therefore achieved before entering desert habitats. In Cactaceae, all desert species evolved from lineages of CAM stem succulents from semiarid habitats (Gibson and Nobel 1986), and adaptations specifically for desert conditions most likely involved greater tolerance to temperature and water stresses for adults and seedlings. Leaf-succulent *Aloe* and Crassulaceae occur within deserts but first underwent great diversification in semiarid habitats. Among Bromeliaceae, water-absorbing trichomes (Fig. 6.6) are found within the family, but have been exploited by desert taxa to function as epiphytes using fog moisture. The tribe Stapelieae originated from leafy plants like the Asian *Frerea indica*, which has relatively thin CAM stems and C_3 leaves (Lange and Zuber 1977), whereas the functionally aphyllous CAM succulent taxa in the tribe have radiated in semiarid and arid Arabia, Africa, and offshore islands.

7.2 Warm Desert Versus Semiarid Habitats

7.2.1 Leaves

For most warm-desert perennials, as well as annuals, either microphylly or aphylly is typical for desert sites where soils are dry most of the year, and for the majority of species, leaves are drought-deciduous. In contrast, sclerophyllous scrub (maquis, chaparral, and mattoral) of mediterranean-type climates is a fire-adapted vegetation that, on average, has substantially larger and wider, often tougher, leaves as compared with desert species, and most habitats are dominated by evergreens. Nevertheless, when authors have made generalizations about "xerophytes", i.e., plants from dry habitats (Chap. 1), examples from semiarid sclerophylls have been heavily cited (e.g., Oppenheimer 1960; Fahn and Cutler 1992). In fact, leaves of woody plants from these two biomes exhibit few anatomical similarities (Table 7.2).

Table 7.2. Anatomical features of leaves from the most common evergreen (E) and drought-decicuous (D), nonsucculent woody dicotyledons of coastal chaparral (42 spp.) versus desert scrub (55 spp.) from southern California and western Baja California, Mexico, Number of species having each feature is noted

	Chaparral		Desert scrub	
	E (29 spp.)	D (13 spp.)	E (12 spp.)	D (43 spp.)
Anatomical feature				
Mean leaf thickness				
<200 μm		6	2	7
200–400 μm	15	7	5	22
>400 μm	14			
Dense nonglandular trichomes				
adaxial	0	4	3	12
abaxial	8	7	4	14
Cutinized outer epidermal wall >5 μm	21	4	10	17
Stomatal features				
amphistomatry	8	9	8	42
hypostomaty	21	4	4	1
stomatal crypts	5	0	1	0
Hypodermis or multiple epidermis	6	0	3	0
Mesophyll				
bifacial	19	7	5	1
isolateral to centric	10	6	7	42
Conspicuous vein sclerenchyma	16	1	6	5

Evergreen leaves of sclerophyllous scrub, which endure summer drought stress, are tough because they possess thick cell walls in adaxial epidermis, often form a hypodermis, show a high degree of mesophyll cell contact, and tend to have lignin-rich sclerenchyma associated with vascular bundles. These cell wall features, regarded as mechanisms to resist tissue damage due to wilting, yield higher dry biomass and cost more to make and maintain than drought-deciduous leaves (Specht 1969; Merino et al. 1984; Mooney 1989). Moreover, the sclerophyll leaf design is generally interpreted as a response to poor soil nutrients (Beadle 1966), and such evergreens have thick mesophyll but low photosynthetic rates (Mooney 1981, 1983). Most evergreen species have hypostomatic leaves, but amphistomaty occurs in each community (Kummerow 1973; Riveros et al. 1976). Often, stomata are hidden on the abaxial side with deep chambers, stomatal crypts, abaxial stomatal furrows, or revolute margins, and they have a well-cutinized adaxial surface. Sclerophyllous, fixed erectophilic, evergreen leaves tend to be amphistomatic (e.g., *Arctostaphylos*), whereas deciduous shrubs generally have hypostomatic leaves (Hoffmann 1972). In mediterranean-type vegetation, canopies of shrubs are too dense for high PPFD to reach most interior leaves.

In sharp contrast are the full-sun leaves of deciduous warm desert shrubs, described in Chapter 3. With rare exceptions, virtually all woody perennials inhabiting warm deserts, including deciduous and evergreen life forms, possess amphistomatic leaves with isolateral mesophyll and high A^{mes}/A (Table 7.2). This is a certain case of convergent evolution, and interpreted as a design for plants living under conditions of high solar irradiance to maximize CO_2 uptake whenever stomata can remain open (Mott et al. 1982). Thick cell walls, if present at all, occur mainly in the epidermis, and sclerenchyma is not common and rarely abundant. The relatively few species of evergreen shrubs of warm desert tend to have thickened epidermal cell walls, and a few have sunken stomata, hypodermis, or a high degree of mesophyll cell contact. The only hypostomatic sclerophyllous desert leaves are present on those few species that probably colonized deserts from sclerophyllous communities, e.g., Rhamnaceae, Rosaceae, Thymelaeaceae, and Berberidaceae, which generally do not occur in the most arid desert sites. Plants with hypostomatic desert leaves are probably not successful colonizers of warm desert habitats.

Among geophytes in Iphigenieae (Colchicaceae), the deserticolous *Hexacyrtis dickeana* is the only amphistomatic representative and has the thickest epidermal and hypodermal walls (Baijnath 1988).

High alpine herbaceous perennials, living under conditions of strong solar radiation and low temperature, have thick, narrow leaves with well-developed palisade mesophyll, but, unlike perennials of full-sun lowland desert scrub, tend to have few adaxial stomata (Körner et al. 1989), and would be expected to have much lower maximum rates of photosynthesis for leaves of equal thickness.

Herbaceous perennials of coastal sand dunes, living under conditions of strong solar radiation but moderate temperature and high levels of salt spray, have amphistomatic, isolateral leaves that are similar to leaves of desert perennials but exhibit higher degrees of fleshiness as a response to salt stress (Purer 1936).

Unlike warm deserts, in woody perennials of a cool desert hypostomaty dominates (Pyykko 1966).

No broad comparative data are available for tropical thorn scrub, which is expected to be the most similar to desert leaves for full-sun plants. Fouquieriaceae, mostly occurring in tropical semiarid habitats, represent a case of preadaptation for having wandlike branches with long shoot-short shoot organization; however, desert species have amphistomatic leaves (Fig. 5.12A) and therefore have evolved structurally and physiologically from relatives in semiarid habitats (Henrickson 1972).

7.2.2 Wood

Several recent studies have compared the functional wood anatomy of desert scrub versus semiarid habitats, particularly chaparral (Carlquist 1985, 1988; Carlquist and Hoekman 1985; Fahn et al. 1986). Nonsucculent chaparral shrubs, like most desert shrubs (Chap. 6), have stem woods with high numbers of extremely narrow vessels, short to extremely short vessel elements with simple perforation plates, and relatively high values of vessel grouping with vasicentric tracheids. Such features, rare in mesic tropical and subtropical habitats, demonstrate that woods of arid and semiarid habitats are highly specialized (Carlquist 1975). Carlquist (1989) has interpreted the design of dryland woods as an adaptive strategy for "safety" to prevent total xylem dysfunction during drought, when air emboli would form in stems that experience extremely low water potentials (Chap. 6). Some extremely low summer water potentials (< -7 MPa) have been measured in certain species of Californian chaparral (Comstock and Mahall 1985; Davis 1989; Vankat 1989). Tertiary helical thickenings, which are commonly observed in these woods, were hypothesized as aids to maintain integrity of the hydraulic system during dry or cold periods (Carlquist 1988).

If preserving the safety of hydraulic function has been a key issue in wood evolution for dry habitats, one expects that desert evergreens, which transpire daily at physiologically significant rates, would be ideal tests, and Carlquist (1980) interpreted xylem design of the desert evergreen *Larrea tridentata* as a good fit. *Larrea* has summer midday shoot water potentials < -8 MPa, but, contrary to expectations, this dominant species has wood with solitary vessels, often at relatively low densities and without tertiary helical thickenings, and imperforate tracheary elements that appear to be too thick-walled to function as conducting cells. Critical studies on the hydraulic systems of evergreens from both habitat types are needed to determine whether physiologically significant water flow occurs in such wood tracheids.

7.2.3 Ontogeny of Desert Shoots

Whereas shoot ontogeny has been studied in several families of succulents (Chap. 5) and aphyllous, green-stemmed shrubs (Chap. 4), in general, the developmental anatomy of shoots has not been investigated in adequate detail for other representative nonsucculent perennials and annuals in any warm desert.

Studies from water-stress experiments and sun versus shade leaves have been used to show that desert leaves have relatively little cell enlargement, which would in part account for high A^{mes}/A ratios and small intercellular air spaces in mesophyll (Turrell 1936; Shields 1951a). Desert species need to be used for ontogenetic studies, especially to determine how different regimes of PPFD and conditions of shoot water potential influence cell division patterns of mesophyll.

One expects that microphylly resulted by decreasing the period of plate meristem, or marginal meristem activity, or both, and that inconspicuous leaf veins would be attributed to very limited activity of adaxial meristem, which in mesophytes typically results in a buildup of cells for principal procambial strands.

Several shoot designs are common within deserts. For long shoot-short shoot development (Fig. 1.5), studies testing effects of growth substances are needed to interpret the lack of internodal formation on short shoots (Gibson 1981). Only through ontogenetic studies will we begin to understand the mechanisms of how a leafy shoot has evolved into a sclerified, aphyllous photosynthetic shoot (Böcher and Lyshede 1968, 1972; Gibson 1979, 1983) and then begin to assess the costs and benefits of each design. Because leaf size and hormonal concentrations are relevant to differentiation patterns of xylem vessels, elucidating the adaptive strategies of the hydraulic systems will require three-dimensional structural analyses of entire shoots integrated with experiments on shoot water relations.

7.3 Future Directions of Structure-Function Research

Thorough cladistic analyses, genus by genus, and for each desert, should be a goal to pinpoint the sister taxon to each desert endemic. Beginning with this knowledge, some very interesting comparative ecophysiology, experimental physiological anatomy, and comparative developmental anatomy can be pursued that will determine how and why desert adaptations arose. With the exception of the genus *Encelia* (Ehleringer et al. 1976, 1981), modern comparative physiological investigations have not been published comparing true sister species, one living in a semiarid habitat and the other that evolved in an arid habitat. For desert plant biology, this is the recipe for a future subdiscipline, but it requires a special research effort to investigate plant ecophysiology in the dry tropics, where many of the ancestors of the desert flora reside.

References

Ackerman TL, Romney EM, Wallace A, Kinnear JE (1980) Phenology of desert shrubs in southern Nye County, Nevada. Great Basin Nat Mem (4): 4–23

Adams MS, Strain BR (1968) Photosynthesis in stems and leaves of *Cercidium floridum*. Spring and summer diurnal field response and relation to temperature. Oecol Plant 3: 285–297

Adams MS, Strain BR (1969) Seasonal photosynthetic rates in stems of *Cercidium floridum* Benth. Photosynthetica 3: 55–62

Adams MS, Strain BR, Ting IP (1967) Photosynthesis in chlorophyllous stem tissue and leaves of *Cercidium floridum*: accumulation and distribution of ^{14}C $^{14}CO_2$. Plant Physiol 42: 1797–1799

Agami M (1987) The establishment, growth, phenology and longevity of two species of *Helianthemum* in the Negev Desert highlands, Israel. Oecologia 71: 486–490

Albert R (1975) Salt regulation in halophytes. Oecologia 21: 55–71

Althawadi AM, Grace J (1986) Water use by the desert cucurbit *Citrullus colocynthis* (L.) Schrad. Oecologia 70: 475–480

Amiran DHK, Wilson AW (eds) (1973) Coastal deserts. Their nature and human environments. University of Arizona Press, Tucson

Appleby RF, Davies WJ (1983) A possible evaporation site in the guard cell wall and the influence of leaf structure on the humidity response by stomata of woody plants. Oecologia 56: 30–40

Arroyo S (1986) Leaf anatomy in the Tecophilaeaceae. J Linn Soc Bot 93: 323–328

Ashby E (1932) Transpiratory organs of *Larrea tridentata* and their ecological significance. Ecology 13: 182–188

Baas P, Carlquist S (1985) A comparison of the ecological wood anatomy of the floras of southern California and Israel. Int Assoc Wood Anat Bull New Ser 6: 349–353

Baas P, Werker E, Fahn A (1983) Some ecological trends in vessel characters. Int Assoc Wood Anat Bull New Ser 4: 141–159

Baijnath H (1988) A contribution to the leaf anatomy of some southern African Iphigenieae (Colchicaceae). S Afr J Bot 54: 265–272

Bailey DC (1980) Anomalous growth and vegetative anatomy of *Simmondsia chinensis*. Am J Bot 67: 147–161

Balding FR, Cunningham GL (1976) A comparison of heat transfer characteristics of simple and pinnate leaf models. Bot Gaz 137: 65–74

Barbour MG, Major J (eds) (1977) Terrestrial vegetation of California. John Wiley, New York

Barcikowski W, Nobel PS (1984) Water relations of cacti during desiccation: distribution of water to tissues. Bot Gaz 145: 110–115

Batanouny KH, Abdel Waheb AM (1973) Eco-physiological studies on desert plants VIII. Root penetration of *Leptadenia pyrotechnica* (Forsk.) Decne. in relation to its water balance. Oecologia 11: 151–161

Beadle NCW (1966) Soil phosphate and its role in molding segments of the Australian flora and vegetation, with special reference to xeromorphy and sclerophylly. Ecology 47: 992–1007

Beal WJ (1886) The bulliform or hygroscopic cells of grasses and sedges compared. Bot Gaz 12: 321–326

Beard JS (1981) Vegetation of Central Australia. In: Jessop J (ed) Flora of Central Australia. Australian Systematic Botany Society, Reed Books, Sydney, pp xxi–xxvi

Beatley JC (1967) Survival of winter annuals in the northern Mojave Desert. Ecology 48: 745–750

Beatley JC (1970) Perennation in *Astragalus lentiginosus* and *Tridens pulchellum* in relation to rainfall. Madroño 20: 326–332

Beatley JC (1974) Phenological events and their environmental triggers in Mojave Desert ecosystems. Ecology 55: 856–863

Beaumont J, Cutler DF, Reynolds T, Vaughan JG (1985) The secretory tissue of aloes and their allies. Isr J Bot 34: 265–282

Becker DA (1968) Stem abscission in the tumbleweed, *Psoralea*. Am J Bot 55: 753–756

Becker DA (1978) Stem abscission in tumbleweeds of the Chenopodiaceae: *Kochia*. Am J Bot 65: 375–383

Begg JE (1980) Morphological adaptations of leaves to water stress. In: Turner NC, Kramer PJ (eds) Adaptation of plants to water and high temperature stress. Wiley–Interscience, New York, pp 33–42

Bemis WP, Whitaker TW (1969) The xerophytic *Cucurbita* of northwestern Mexico and southwestern United States. Madroño 20: 33–41

Bennert HW, Mooney HA (1979) The water relations of some desert plants in Death Valley, California. Flora 168: 405–427

Bennert HW, Schmidt B (1983) Untersuchungen zur Salzabscheidung bei *Atriplex hymenelytra* (Torr.) Wats. (Chenopodiaceae). Flora 174: 341–355

Benzing DH (1970) Foliar permeability and the absorption of minerals and organic nitrogen by certain tank bromeliads. Bot Gaz 131: 23–31

Benzing DH (1980) The biology of the bromeliads. Mad River Press, Eureka

Benzing DH (1990) Vascular epiphytes. Cambridge University Press, Cambridge

Bergstrom G, Schaller M, Eickmeier WG (1982) Ultrastructural and biochemical bases of resurrection in the drought-tolerant vascular plant, *Selaginella lepidophylla*. J Ultrastruct Res 78: 269–282

Berry J, Björkman O (1980) Photosynthesis response and adaptation to temperature in higher plants. Annu Rev Plant Physiol 31: 491–543

Bessey CE (1904) The chimney-shaped stomata of *Holacantha emoryi*. Bull Torrey Bot Club 31: 523–527

Bewley JD (1979) Physiological aspects of desiccation tolerance. Annu Rev Plant Physiol 30: 195–238

Bhandari MM (1978) Flora of the Indian Desert. Scientific Publishers, Jodhpur

Björkman O (1981) Responses to different quantum flux densities. In: Lange OL, Nobel PS, Osmond CB, Ziegler H (eds) Encyclopedia of plant physiology, new ser, vol 12A. Physiological plant ecology I. Responses to the physical environment. Springer, Berlin Heidelberg New York, pp 57–107

Björkman O, Badger MR, Armond PA (1980) Response and adaptation of photosynthesis to high temperature. In: Turner NC, Kramer PJ (eds) Adaptations of plants to water and high temperature stress. Wiley-Interscience, New York, pp 223–249

Black RF (1954) The leaf anatomy of Australian members of the genus *Atriplex*. I. *Atriplex vesicaria* Heward and *A. nummularia* Lindl. Aust J Bot 2: 269–286

Black RF (1958) Effect of sodium chloride on leaf succulence and area of *Atriplex hastata* L. Aust J Bot 6: 306–321

Bleckmann CA, Hull HW (1975) Leaf and cotyledon surface ultrastructure of five *Prosopis* species. J Ariz Acad Sci 10: 98–105

Blunden G, Yi Y, Jewers K (1973) The comparative leaf anatomy of *Agave, Beschorneria, Doryanthes* and *Furcraea* species (Agavaceae: Agaveae). J Linn Soc Bot 66: 157–179

Böcher TW (1971) Anatomical studies in cottonthorn, *Tetradymia axillaris* A. Nels. Nat Can 98: 225–250

Böcher TW (1972) Comparative anatomy of three of the apophyllous genus *Gymnophyton*. Am J Bot 59: 494–503

Böcher TW (1975) Structure of the multinodal photosynthetic thorns in *Prosopis kuntzei* Harms. Biol Skrift K Dan Vidensk Selsk 20(8): 1–43

Böcher TW (1979) Xeromorphic leaf types. Evolutionary strategies and tentative semophyletic sequences. Biol Skrift K Dan Vidensk Selsk 23(8): 1–71

Böcher TW, Lyshede OB (1968) Anatomical studies in xerophytic apophyllous plants: I. *Monttea aphylla, Bulnesia retama* and *Bredemeyera colletioides*. Biol Skrift K Dan Vidensk Selsk 16(3): 1–44

Böcher TW, Lyshede OB (1972) Anatomical studies in xerophytic apophyllous plants. II. Additional species from South American shrub steppes. Biol Skrift K Dan Vidensk Selsk 18(4): 1–137

Boeken B (1990) Life histories of two desert species of the bulbous genus *Bellevalia*. The relation between biomass partitioning and water availability. Oecologia 82: 172–179

Bornkamm R, Kehl H (1989) Landscape ecology of the western desert of Egypt. J Arid Environ 17: 271–277

Bosabalidis AM, Thomson WW (1984) Ultrastructural differentiation of an unusual structure lining the anticlinal walls of the inner secretory cells in *Tamarix* salt glands. Bot Gaz 145: 427–435

Bossard CC, Rejmanek M (1992) Why have green stems? Funct Ecol 6: 197–205

Boughton VH (1986) Phyllode structure, taxonomy and distribution in some Australian acacias. Aust J Bot 34: 663–674

Boughton VH (1990) Aspects of phyllode anatomy in some Australian phyllodinous acacias, with particular regard to stickiness. Aust J Bot 38: 131–151

Box EO (1981) Macroclimate and plant forms: an introduction to predictive modeling in phytogeography. Dr W Junk, The Hague

Brandham PE, Cutler DF (1978) Influence of chromosome variation on the organisation of the leaf epidermis in a hybrid *Aloe* (Liliaceae). J Linn Soc Bot 77: 1–16

Bray WL (1903) The tissues of some of the plants of the Sotol Region. Bull Torrey Bot Club 30: 621–633

Brown DE (ed) (1982) Biotic communities of the American Southwest—United States and Mexico. Desert Plants 4: 1–341

Brown JH, Gibson AC (1983) Biogeography. CV Mosby, St Louis

Bunce JA (1982) Low humidity effects of photosynthesis in single leaves of C_4 plants. Oecologia 54: 233–235

Bunce JA (1983) Differential sensitivity to humidity of daily photosynthesis in the field in C_3 and C_4 species. Oecologia 57: 262–265

Bunce JA (1985) Effect of boundary layer conductance on the response of stomata to humidity. Plant Cell Environ 8: 55–57

Bunce JA (1986) Responses of gas exchange to humidity in populations of three herbs from environments differing in atmospheric water. Oecologia 71: 117–120

Burbridge NT (1946) Foliar anatomy and the delimitation of the genus *Triodia* R. Br. Blumea 3(Suppl): 83–89

Burgess TL (1985) Agave adaptation to aridity. Desert Plants 7: 39–50

Butler V, Bornman CH, Evert RF (1973) *Welwitschia mirabilis:* morphology of the seedling. Bot Gaz 134: 52–59

Buxbaum F (1950) Morphology of cacti. Section I. Roots and stems. Abbey Garden Press, Pasadena

Caldwell MM (1974) Physiology of desert halophytes. In: Reimold RJ, Queen WH (eds) Ecology of halophytes. Academic Press, New York, pp 355–378

Caldwell MM (1981) Plant response to solar ultraviolet radiation. In: Lange OL, Nobel PS, Osmond CB, Ziegler H (eds) Encyclopedia of plant physiology, new ser, vol 12A. Physiological plant ecology I. Responses to the physical environment. Springer, Berlin Heidelberg New York, pp 169–197

Calkin HW, Pearcy RW (1984) Seasonal progressions of tissue and cell water relations parameters in evergreen and deciduous perennials. Plant Cell Environ 7: 347–352

Calvin CL (1970) Anatomy of the aerial epidermis of the mistletoe, *Phoradendron flavescens*. Bot Gaz 131: 62–74

Cannell MGR (1982) World forest biomes and primary production data. Academic Press, London

Cannon WA (1908) The topography of the chlorophyll apparatus in desert plants. Carnegie Inst Wash Pub 98, Washington, DC

Cannon WA (1911) Root habits of desert plants. Carnegie Inst Wash Pub 131, Washington, DC

Cannon WA (1924) General and physiological features of the vegetation of the more arid portions of southern Africa, with notes on the climatic environment. Carnegie Inst Wash Pub 354, Washington, DC

Carlquist S (1962) A theory of paedomorphosis in dicotyledonous woods. Phytomorphology 12: 30–45

Carlquist S (1975) Ecological strategies of xylem evolution. University of California Press, Berkeley
Carlquist S (1980) Further concepts in ecological wood anatomy, with comments on recent work in wood anatomy and evolution. Aliso 9: 499–553
Carlquist S (1982) Wood anatomy of Buxaceae: correlations with ecology and phylogeny. Flora 172: 463–491
Carlquist S (1984) Vessel grouping in dicotyledon wood: significance and relationship to imperforate tracheary elements. Aliso 10: 505–525
Carlquist S (1985) Vasicentric tracheids as a drought survival mechanism in the flora of southern California and similar regions. Aliso 11: 37–86
Carlquist S (1988) Near-vessellessness in *Ephedra* and its significance. Am J Bot 75: 598–601
Carlquist S (1989) Adaptive wood anatomy of chaparral shrubs. In: Keeley JE (ed) The California chaparral: paradigms reexamined. Natural History Museum of Los Angeles County Science Series 34, Los Angeles, pp 25–35
Carlquist S, Hoekman DA (1985) Ecological wood anatomy of the woody southern California flora. Int Assoc Wood Anat Bull New Ser 6: 319–347
Carolin RC, Jacobs SWL, Vesk M (1975) Leaf structure in Chenopodiaceae. Bot Jahrb Syst 95: 226–255
Carolin RC, Jacobs SWL, Vesk M (1978) Kranz cells and mesophyll in Chenopodiaceae. Aust J Bot 26: 683–698
Carolin RC, Jacobs SWL, Vesk M (1982) The chlorenchyma of some members of the Salicornieae (Chenopodiaceae). Aust J Bot 30: 387–392
Carrodus BB, Specht RL (1965) Factors affecting the relative distribution of *Atriplex vesicaria* and *Kockia sedifolia* (Chenopodiaceae) in the arid zone of South Australia. Aust J Bot 13: 419–433
Cavagnaro JB (1988) Distribution of C_3 and C_4 grasses at different altitudes in a temperate arid region of Argentina. Oecologia 76: 273–277
Chabot BF, Chabot JF (1977) Effects of light and temperature on leaf anatomy and photosynthesis in *Fragaria vesca*. Oecologia 26: 363–373
Chabot BF, Jurik TW, Chabot JF (1979) Influence of instantaneous and integrated light-flux density on leaf anatomy and photosynthesis. Am J Bot 66: 940–945
Chamberlain CJ (1935) Gymnosperms. Structure and evolution. University of Chicago Press, Chicago
Cheeseman JM (1988) Mechanisms of salinity tolerance in plants. Plant Physiol 87: 547–550
Chew RM, Chew AE (1965) Primary productivity of a desert shrub (*Larrea tridentata*) community. Ecol Monogr 35: 355–375
Christodoulakis NS (1989) An anatomical study of seasonal dimorphism in the leaves of *Phlomis fruticosa*. Ann Bot 63: 389–394
Cloudsley-Thompson J (1977) Man and the biology of arid zones. Edward Arnold, London
Cody ML (1978) Distribution ecology of *Haplopappus* and *Chrysothamnus* in the Mojave Desert. I. Niche position and niche shifts on north-facing granitic slopes. Am J Bot 65: 1107–1116
Cole DT (1987) *Lithops* of SWA/Namibia. Madoqua Mem (Windhoek) 1
Comstock J, Ehleringer J (1984) Photosynthetic responses to slowly decreasing leaf water potentials in *Encelia frutescens*. Oecologia 61: 241–248
Comstock J, Ehleringer J (1986) Canopy dynamics and carbon gain in response to soil water availability in *Encelia frutescens* Gray, a drought-deciduous shrub. Oecologia 68: 271–278
Comstock JP, Ehleringer JR (1988) Contrasting photosynthetic behavior in leaves and twigs of *Hymenoclea salsola*, a green-twigged warm desert shrub. Am J Bot 75: 1360–1370
Comstock JP, Mahall BE (1985) Drought and changes in leaf orientation for two California chaparral shrubs: *Ceanothus megacarpus* and *Ceanothus crassifolius*. Oecologia 65: 531–535
Comstock JP, Cooper TA, Ehleringer JR (1988) Seasonal patterns of canopy development and carbon gain in nineteen warm desert shrub species. Oecologia 75: 327–335
Cowan IR (1977) Stomatal behavior and environment. Adv Bot Res 4: 117–227
Cowan IR (1981) Coping with water stress. In: Pate J, McCoomb AJ (eds) Biology of Australian native plants. University of Western Australia Press, Perth, pp 1–32
Cowan IR, Farquhar GD (1977) Stomatal function in relation to leaf metabolism and environment. Symp Soc Exp Bot 31: 471–505
Cunningham GL, Strain BR (1968) Irradiance and productivity in a desert shrub. Photosynthetica 3: 69–71

Cunningham GL, Strain BR (1969) Ecological significance of seasonal leaf variability in a desert shrub. Ecology 50: 400–408

Cutler DF (1972) Leaf anatomy of certain *Aloe* and *Gasteria* species and their hybrids. In: Ghouse AKM, Yunus M (eds) Research trends in plant anatomy. Tata McGraw Hill, New Delhi, pp 103–122

Cutler DF, Rains DM, Loomis RS (1977) The importance of cell size in the water relations of plants. Physiol Plant 40: 255–260

Danin A (1983) Desert vegetation of Israel and Sinai. Cana Publishing House, Jerusalem

Darling MS (1989) Epidermis and hypodermis of the saguaro cactus (*Cereus giganteus*): anatomy and spectral properties. Am J Bot 76: 1698–1706

Darrow RA (1943) Vegetative and floral growth of *Fouquieria splendens*. Ecology 24: 310–322

Davis SD (1989) Patterns in mixed chaparral stands—differential water status and seedling survival during summer drought. In: Keeley JE (ed) The California chaparral: paradigms reexamined. Natural History Museaum of Los Angeles Country Science Series 34, Los Angeles, pp 97–105

Delf EM (1912) Transpiration in succulent plants. Ann Bot 26: 409–442

Delf EM (1915) The meaning of xerophily. J Ecol 3: 110–121

Dell B, McCoomb AJ (1978) Plant resins—their formation, secretion and possible functions. In: Woolhouse EW (ed) Advances in botanical research. Academic Press, London, pp 277–316

Dengler NG, Dengler RF, Hattersley PW (1985) Differing ontogenetic origins of PCR ("Kranz") sheaths in leaf blades of C_4 grasses (Poaceae). Am J Bot 72: 284–302

DePuit EJ, Caldwell MM (1975) Stem and leaf gas exchange of two arid land shrubs. Am J Bot 62: 954–961

De Santo AV, Alfani A, De Luca P (1976) Water vapour uptake from the atmosphere by some *Tillandsia* species. Ann Bot 40: 391–394

de Winter B (1965) The South African Stipeae and Aristideae (Gramineae). Bothalia 8: 201–404

Didden-Zopfy B, Nobel PS (1982) High temperature tolerance and heat acclimation of *Opuntia bigelovii*. Oecologia 52: 176–180

Dilcher DL, Christophel DC, Bhagwandin HO Jr, Scriven LJ (1990) Evolution of the Casuarinaceae: morphological comparisons of some extant species. Am J Bot 77: 338–355

Dittmer HJ (1959) A study of the root systems of certain sand dune plants in New Mexico. Ecology 40: 265–273

Dittmer HJ, Roser ML (1963) The periderm of certain members of the Cucurbitaceae. Southwest Nat 8: 1–9

Dittmer HJ, Talley BP (1964) Gross morphology of tap roots of desert cucurbits. Bot Gaz 125: 121–126

Dörn RI, Oberlander TM (1981) Rock varnish origin, characteristics, and usage. Z Geomorph 25: 420–436

Duerden H (1934) On the occurrence of vessels in *Selaginella*. Ann Bot Ser 1 48: 459–465

Ehleringer JR (1976) Leaf absorptance and photosynthesis as affected by pubescence in genus *Encelia*. Carnegie Inst Wash Yearbook 75: 413–418

Ehleringer JR (1977) Adaptive value of leaf hairs in *Encelia farinosa*. Carnegie Inst Wash Yearbook 76: 413–418

Ehleringer JR (1980) Leaf morphology and reflectance in relation to water and temperature stress. In: Turner NC, Kramer PJ (eds) Adaptations of plants to water and high temperature stress. Wiley-Interscience, New York, pp 295–308

Ehleringer J (1981a) Ecology and ecophysiology of leaf pubescence in North American desert plants. In: Mabry T, Rodriquez E, Healey P (eds) Plant trichomes. Plenum Press, New York, pp 113–132

Ehleringer J (1981b) Leaf absorptances of Mohave and Sonoran Desert plants. Oecologia 49: 366–370

Ehleringer J (1982) The influence of water stress and temperature on leaf pubescence in *Encelia farinosa*. Am J Bot 69: 670–675

Ehleringer J (1983a) Characterization of a glabrate *Encelia farinosa* mutant: morphology, ecophysiology, and field observations. Oecologia 57: 303–310

Ehleringer J (1983b) Ecophysiology of *Amaranthus palmeri*, a Sonoran Desert summer annual. Oecologia 57: 107–112

Ehleringer J (1985a) Comparative microclimatology and plant responses in *Encelia* species from contrasting habitats. J Arid Environ 8: 45–56
Ehleringer JR (1985b) Annuals and perennials of warm deserts. In: Chabot BF, Mooney HA (eds) Physiological ecology of North American plant communities. Chapman and Hall, New York, pp 162–180
Ehleringer JR (1988) Comparative ecophysiology of *Encelia farinosa* and *Encelia frutescens*. Oecologia 76: 553–561
Ehleringer JR (1994) Variation in gas exchange characteristics among desert plants. In: Schulze E-D, Caldwell MM (eds) Ecophysiology of photosynthesis. Springer, Berlin Heidelberg New York, pp 361–392
Ehleringer JR, Björkman O (1978a) A comparison of photosynthetic characteristics of *Encelia* species possessing glabrous and pubescent leaves. Plant Physiol 62: 185–190
Ehleringer JR, Björkman O (1978b) Pubescence and leaf spectral characteristics in a desert shrub, *Encelia farinosa*. Oecologia 36: 151–162
Ehleringer JR, Cook CS (1987) Leaf hairs in *Encelia* (Asteraceae). Am J Bot 74: 1532–1540
Ehleringer JR, Cook CS (1990) Characteristics of *Encelia* species differing in leaf reflectance and transpiration rate under common garden conditions. Oecologia 82: 484–489
Ehleringer JR, Cooper TA (1988) Correlations between carbon isotope ratio and microhabitat in desert plants. Oecologia 76: 562–566
Ehleringer JR, Cooper TA (1992) On the role of orientation in reducing photoinhibitory damage in photosynthetic-twig desert shrubs. Plant Cell Environ 15: 301–306
Ehleringer JR, Forseth I (1980) Solar tracking by plants. Science 210: 1094–1098
Ehleringer JR, Mooney HA (1978) Leaf hairs: effects on physiological activity and adaptive value to a desert shrub. Oecologia 37: 183–200
Ehleringer JR, Schulze E-D (1985) Mineral concentration in an autoparasitic *Phoradendron californicum* growing on a parasitic *P. californicum* and its host, *Cercidium floridum*. Am J Bot 72: 568–571
Ehleringer JR, Werk KS (1986) Modifications of solar-radiation absorption patterns and implications for carbon gain at the leaf level. In: Givnish TJ (ed) On the economy of plant form and function. Cambridge University Press, Cambridge, pp 57–82
Ehleringer JR, Björkman, O Mooney HA (1976) Leaf pubescence: effects on absorptance and photosynthesis in a desert shrub. Science 192: 376–377
Ehleringer J, Mooney HA, Gulmon SL, Rundel P (1980) Orientation and its consequences for *Copiapoa* (Cactaceae) in the Atacama Desert. Oecologia 46: 63–67
Ehleringer JR, Mooney HA, Gulmon SL, Rundel PW (1981) Parallel evolution of leaf pubescence in *Encelia* in coastal deserts of North and South America. Oecologia 49: 38–41
Ehleringer JR, Schulze E-D, Ziegler H, Lange OL, Farquhar GD, Cowan IR (1985) Xylem-tapping mistletoes: water or nutrient parasites? Science 227: 1479–1481
Ehleringer JR, Ullmann I, Lange OL, Farquhar GD, Cowan IR, Schulze E-D, Ziegler H (1986) Mistletoes: a hypothesis concerning morphological and chemical avoidance of herbivory. Oecologia 70: 234–237
Ehleringer JR, Comstock JP, Cooper TA (1987) Leaf-twig carbon isotope ratio differences in photosynthesizing-twig desert shrubs. Oecologia 71: 318–320
Eickmeier WG (1978) Photosynthetic pathway distributions along an aridity gradient in Big Bend National Park, and implications for enhanced resource partitioning. Photosynthetica 12: 290–297
Eickmeier WG (1979) Photosynthetic recovery in the resurrection plant *Selaginella lepidophylla* after wetting. Oecologia 39: 93–106
Eickmeier WG (1980) Photosynthetic recovery of resurrection spikemosses from different hydration regimes. Oecologia 46: 380–385
Eickmeier WG (1982) Protein synthesis and photosynthetic recovery in the resurrection plant, *Selaginella lepidophylla*. Plant Physiol 69: 135–138
Eickmeier WG (1983) Photosynthetic recovery of the resurrection plant *Selaginella lepidophylla* (Hook. and Grev.) Spring: effects of prior desiccation rate and mechanisms of desiccation damage. Oecologia 58: 115–120
Ellenberg H (1959) Über den Wasserhaushalt tropischer Nebeloasen in der Küstenwüste Perus. Berl Geobot Forsch Inst Rübel 1958: 47–74

Ellenberg H (1981) Ursachen des Vorkommens und Fehlens von Sukkulenten in den Trockengebieten der Erde. Flora 171: 114–169
Eller BM, Grobbelaar N (1982) Geophylly: consequences for *Ledebouria ovatifolia* in its natural habitat. J Exp Bot 33: 366–375
Eller BM, Grobbelaar N (1986) Diurnal temperature variation in and around a *Lithops lesliei* plant growing in its natural habitat on a clear day. S Afr J Bot 52: 403–407
Eller BM, Nipkow A (1983) Diurnal course of the temperature in a *Lithops* sp. (Mesembryanthemaceae Fenzl) and its surrounding soil. Plant Cell Environ 6: 559–565
Ellis RP (1976) A procedure for standardizing comparative leaf anatomy in the Poaceae. I. The leaf-blade as viewed in transverse section. Bothalia 12: 65–109
Ellis RP (1977) Leaf anatomy of the South African Danthonieae (Poaceae). I. The genus *Dregochloa*. Bothalia 12: 209–213
Ellis RP (1984) Leaf anatomy of the South African Danthonieae (Poaceae). IX. *Asthenatherium glaucum*. Bothalia 15: 153–159
El-Monayeri MO, Youssef MM, El-Ghamry AA (1981) Contributions to the autecology of two *Zygophyllum* species growing in the Egyptian Desert. Egypt J Bot 24: 49–68
Emerson FW (1935) An ecological reconnaissance in the White Sands, New Mexico. Ecology 16: 226–233
Evenari M (1938a) Root conditions of certain plants in the wilderness of Judaea. J Linn Soc Bot 51: 383–388
Evenari M (1938b) The physiological anatomy of the transpiratory organs and the conducting systems of certain plants typical of the wilderness of Judaea. J Linn Soc Bot 51: 389–407
Evenari M (1985a) The desert environment. In: Evenari M, Noy-Meir I, Goodall DW (eds) Hot deserts and arid shrublands. Ecosystems of the world, vol 12A. Elsevier, Amsterdam, pp 1–22
Evenari M (1985b) Adaptations of plants and animals to the desert, environment. In: Evenari M, Noy-Meir I, Goodall DW (eds) Hot deserts and arid shrublands. Ecosystems of the world, vol 12A. Elsevier, Amsterdam, pp 79–92
Evenari M, Gutterman Y (1973) Some notes on *Salvadora persica* L. in Sinai and its use as a toothbrush. Flora 162: 118–125
Evenari M, Richter R (1938) Physiological-ecological investigations in the Wilderness of Judaea. J Linn Soc Bot 51: 333–381
Evenari M, Shanan L, Tadmor N (1971) The Negev: the challenge of a desert. Harvard University Press, Cambridge
Evenari M, Schulze E-D, Kappen L, Buschbom U, Lange OL (1975) Adaptive mechanisms in desert, plants. In: Vernbern FJ (ed) Physiological adaptation to the environment. Intext Education, New York, pp 111–129
Evenari M, Shanan L, Tadmor N (1982) The Negev: the challenge of a desert, 2nd edn. Harvard University Press, Cambridge
Evenari M, Noy-Meir I, Goodall DW (eds) (1985) Hot deserts and arid shrublands. Ecosystems of the world, vol 12A. Elsevier Science, Amsterdam
Evenari M, Noy-Meir I, Goodall DW (eds) (1986) Hot deserts and arid shrublands. Ecosystems of the world, vol 12B. Elsevier Science, Amsterdam
Ezcurra E, Montaña C, Arizaga S (1991) Architecture, light interception, and distribution of *Larrea* species in the Monte Desert, Argentina. Ecology 72: 23–34
Fahmy GM, Hegazy AK, Kassan HT (1990) Phenology, pigment content and diurnal change of proline in green and senescing leaves of three *Zygophyllum* species. Flora 184: 423–436
Fahn A (1958) Xylem structure and annual rhythm of development in trees and shrubs of the desert: I, *Tamarix aphylla*, T. *jordanis* var. *negevensis*, *T. gallica* var. *maris-mortui*. Trop Woods 109: 81–94
Fahn A (1959) Xylem structure and annual rhythm of development in trees and shrubs of the desert: II, *Acacia tortilis* and *A. raddiana*. Bull Res Counc Isr 7D: 23–28
Fahn A (1963) The fleshy cortex of articulated Chenopodiaceae. Maheshwari Commemorative Volume. J Ind Bot Soc 42A: 39–45
Fahn A (1964) Some anatomical observations of desert plants. Phytomorphology 14: 93–102
Fahn A (1979) Secretory tissues in plants. Academic Press, London
Fahn A (1986) Structure and functional properties of trichomes of xeromorphic leaves. Ann Bot 57: 631–637

Fahn A (1988) Secretory tissues in vascular plants. New Phytol 108: 229–257

Fahn A (1990) Plant anatomy, 4th edn. Pergamon Press, Oxford

Fahn A, Arzee T (1959) Vascularization of articulated Chenopodiaceae and the nature of their fleshy cortex. Am J Bot 46: 330–338

Fahn A, Broido S (1963) The primary vascularization of the stems and leaves of the genera *Salsola* and *Suaeda* (Chenopodiaceae). Phytomorphology 13: 156–165

Fahn A, Cutler DF (1992) Xerophytes. Handbuch der Pflanzenanatomie, Bd 13(3). Borntraeger, Berlin

Fahn A, Dembo N (1964) Structure and development of the epidermis in articulated Chenopodiaceae. Isr J Bot 13: 177–192

Fahn A, Leshem B (1963) Wood fibres with living protoplasts. New Phytol 62: 91–98

Fahn A, Sarnat C (1963) Xylem structure and annual rhythm of development in trees and shrubs of the desert: IV, shrubs. Bull Res Counc Isr 11D: 198–209

Fahn A, Waisel Y, Benjamini L (1968) Cambial activity in *Acacia raddiana* Savi. Ann Bot 32: 677–686

Fahn A, Werker E, Baas P (1986) Wood anatomy and identification of trees and shrubs from Israel and adjacent regions. Israel Academy of Sciences and Humanities, Jerusalem

Farquhar GD, Raschke K (1978) On the resistance to transpiration of the sites of evaporation within the leaf. Plant Physiol 61: 1000–1005

Fearn B (1977) Epidermal structure, stomatal distribution and water-loss in *Lithops marmorate.* Cact Succ J Gr Brit 39:.68–70

Field CB (1991) Ecological scaling of carbon gain to stress and resource availability. In: Mooney HA, Winner WE, Pell EJ (eds) Response of plants to multiple stresses. Academic Press, San Diego, pp 35–65

Fioretto A, Alfani A (1988) Anatomy of succulence and CAM in 15 species of *Senecio.* Bot Gaz 149: 142–152

Fisher SG, Minckley WL (1978) Chemical characteristics of a desert stream in flash flood. J Arid Environ 1: 25–33

Flach BMT (1986) Strahlungangebot, Strahlungsgenuss und Photosynthese nicht ebener Blattflächen in Abhängigkeit vom Tagesgang der Sonne. Thesis, University of Zürich (Abstr)

Flowers TJ, Troke PF, Yeo AR (1977) The mechanism of salt tolerance in halophytes. Annu Rev Plant Physiol 28: 89–121

Foote K, Scheadle M (1976) Physiological characteristics of photosynthesis and respiration by stems of *Populus tremuloides* Michx. Plant Physiol 58: 651–655

Foster JF, Smith WK (1986) Influence of stomatal distribution on transpiration in low-wind environments. Plant Cell Environ 9: 751–759

Franco-Vizcaíno E, Goldstein G, Ting IP (1990) Comparative gas exchange of leaves and bark in three stem succulents of Baja California. Am J Bot 77: 1272–1278

Friedman J, Gunderman N, Ellis M (1978) Water response of the hygrochastic skeletons of the true rose of Jericho (*Anastatica hierochontica* L.). Oecologia 32: 289–301

Friedmann EI, Galun M (1974) Desert algae, lichens, and fungi. In: Brown GW Jr (ed) Desert biology, vol 2. Academic Press, New York, pp 165–212

Fuller WH (1974) Desert soils. In: Brown GW Jr (ed) Desert biology, vol 2. Academic Press, New York, pp 31–101

Gaff DF (1977) Desiccation tolerant vascular plants of southern Africa. Oecologia 31: 95–109

Galil J (1958) Physiological studies on the development of contractile roots in geophytes. Bull Res Counc Isr 6D: 221–236

Gamaley YuV (1984) Leaf morphology and anatomy of Gobi Desert plants. Bot J USSR 69: 569–584

Gates DM (1968) Transpiration and leaf temperature. Annu Rev Plant Physiol 19: 211–238

Gates DM (1970) Physical and physiological properties of plants, pp. 224–252. In: National Academy of Sciences, Remote Sensing, Washington, DC

Gates DM (1980) Biophysical ecology. Springer, Berlin Heidelberg New York

Gates DM, Papian LE (1971) Atlas of energy budgets of plant leaves. Academic Press, London

Gates DM, Aldrefer R, Taylor E (1968) Leaf temperatures of desert plants. Science 159: 994–995

Gedalovich E, Fahn A (1983) Ultrastructure and development of the inactive stomata of *Anabasis articulata.* Am J Bot 70: 88–96

Geller GN, Nobel PS (1984) Cactus ribs: influence of PAR interception and CO_2 uptake. Photosynthetica 18: 482–494

Genkel PA (1946) The resistance of plants to drought and how to increase it. Trudy Timiryazev Inst Plant Physiol 5(1) (in Russian)
Gentry HS (1982) Agaves of continental North America. University of Arizona Press, Tucson
Gentry HS, Sauck JR (1978) The stomatal complex in *Agave*; groups Deserticolae, Campaniflorae, and Umbelliflorae. Proc Calif Acad Sci 41: 371–387
Gibson AC (1973) Comparative anatomy of secondary xylem in Cactoideae (Cactaceae). Biotropica 5: 29–65
Gibson AC (1975) Another look at the cactus research of Irving Widmer Bailey. In: Glass C, Foster R (eds) 1975 Yearbook. Cact Succ J US 47(Suppl): 76–85
Gibson AC (1977) Wood anatomy of opuntias with cylindrical to globular stems. Bot Gaz 138: 334–351
Gibson AC (1978a) Structure of *Pterocactus tuberosus*, a cactus geophyte. Cact Succ J US 50: 41–43
Gibson AC (1978b) Wood anatomy of platyopuntias. Aliso 9: 279–303
Gibson AC (1978c) Architectural designs of wood skeletons in cacti. Cact Succ J Gr Brit 40: 73–80
Gibson AC (1979) Anatomy of *Koeberlinia* and *Canotia* revisited. Madroño 26: 1–12
Gibson AC (1981) Vegetative anatomy of *Pachycormus* (Anacardiaceae). J Linn Soc Bot 83: 273–284
Gibson AC (1982) Anatomy of succulence. In: Ting IP, Gibbs M (eds) Crassulacean acid metabolism. American Society of Plant Physiologists, Rockville, pp 1–17
Gibson AC (1983) Anatomy of photosynthetic old stems of nonsucculent dicotyledons from North American deserts. Bot Gaz 144: 347–362
Gibson AC (1994) Vascular tissues. In: Behnke H-D, Mabry TJ (eds) Caryophyllales. Springer, Berlin Heidelberg New York, pp 45–74
Gibson AC, Horak HE (1978) Systematic anatomy and phylogeny of Mexican columnar cacti. Ann MO Bot Gard 65: 999–1057
Gibson AC, Nobel PS (1986) The cactus primer. Harvard University Press, Cambridge
Gibson AC, Calkin HW, Nobel PS (1984) Xylem anatomy, water flow, and hydraulic conductance in the fern *Cyrtomium falcatum*. Am J Bot 71: 564–574
Gibson AC, Bajaj R, McLaughlin JL, Spencer K (1986) The ever-changing landscape of cactus systematics. Ann MO Bot Gard 73: 532–555
Ginzburg C (1963) Some anatomic features of splitting of desert shrubs. Phytomorphology 13: 92–97
Glen HF, Hardy DS (1986) A method for the non-destructive examination of leaves of Aloe species by SEM. Bothalia 16: 53–55
Goldstein G, Nobel PS (1991) Changes in osmotic pressure and mucilage during low-temperature acclimation of *Opuntia ficus-indica*. Plant Physiol 97: 954–961
Goldstein G, Nobel PS (1994) Water relations and low-temperature acclimation for cactus species varying in freezing tolerance. Plant Physiol 104: 675–681
Goldstein G, Ortega JKE, Nerd A, Nobel PS (1991a) Diel patterns of water potential components for the crassulacean acid metabolism plant *Opuntia ficus-indica* when well-watered or droughted. Plant Physiol 95: 274–280
Goldstein G, Sharifi MR, Kohorn LU, Lighton JRB, Schult L, Rundel PW (1991b) Photosynthesis by inflated pods of a desert shrub, *Isomeris arborea*. Oecologia 85: 396–402
Goudie AS, Pye K (eds) (1985) Chemical sediments and geomorphology. Academic Press, London
Goudie A, Wilkinson J (1977) The warm desert environment. Cambridge University Press, Cambridge
Grantz DA (1990) Plant response to atmospheric humidity. Plant Cell Environ 13: 667–679
Gulmon SL, Rundel PW, Ehleringer JR, Mooney HA (1979) Spatial relationships and competition in a Chilean desert cactus. Oecologia 44: 40–43
Gupta RK (1986) The Thar Desert. In: Evenari M, Noy-Meir I, Goodall DW (eds) Hot deserts and arid shrublands. Ecosystems of the world, vol 12B. Elsevier Science, Amsterdam, pp 55–99
Gutterman Y (1993) Seed germination in desert plants. Springer, Berlin Heidelberg New York
Haberlandt G (1914) Physiological plant anatomy. (Translated from 4th German edn by Montagu Drummond.) Macmillan, London
Halevy G, Orshan G (1972) Ecological studies of *Acacia* species in the Negev and Sinai I. Distribution of *Acacia raddiana*, *A. tortilis* and *A. gerrardii* ssp. *negevensis* as related to environmental factors. Isr J Bot 21: 197–208
Halevy G, Orshan G (1973) Ecological studies of *Acacia* species in the Negev and Sinai II. Phenology of *Acacia raddiana*, *A. tortilis* and *A. gerrardii* ssp. *negevensis*. Isr J Bot 22: 120–138

Hammer S, Hartmann HEK (1990) Shedding light on windows—in *Conophytum.* Cact Succ J US 62: 35–42

Han S, Suzaki T (1981) Studies on the production and consumption of assimilates by trees: bark photosynthesis and respiration in young green stems and branches of *Fagus crenata* and *Quercus acutissima.* J Japn For Soc 63: 242–244

Harris JA, Harrison GJ, Pascoe TA (1930) Osmotic concentration and water relations in the mistletoes, with special reference to the occurrence of *Phoradendron californicum* on *Covillea tridentata.* Ecology 11: 687–702

Harten JB, Eickmeier WG (1987) Comparative desiccation tolerance of three desert pteridophytes: response to long-term desiccation. Am Midl Nat 118: 337–347

Hattersley PW (1983) The distribution of C_3 and C_4 grasses in Australia in relation to climate. Oecologia 57: 113–128

Hattersley PW, Watson L (1976) C_4 grasses: an anatomical criterion for distinguishing between NADP-malic enzyme species and PCK or NAD-malic enzyme species. Aust J Bot 24: 297–308

Heckathorn SA, DeLucia E (1991) Effect of leaf rolling on gas exchange and leaf temperature of *Andropogon gerardii* and *Spartina pectinata.* Bot Gaz 152: 263–268

Henrickson J (1969a) An introduction to the Fouquieriaceae. Cact Succ J US 41: 97–105

Henrickson J (1969b) Anatomy of periderm and cortex of Fouquieriaceae. Aliso 7: 97–126

Henrickson J (1969c) The succulent fouquierias. Cact Succ J US 41: 178–184

Henrickson J (1972) A taxonomic revision of the Fouquieriaceae. Aliso 7: 439–537

Henrickson J (1975) Vegetative morphology of the cirio (*Fouquieria columnaris*). In: Glass C, Foster R (eds) 1975 Yearbook. Cact Succ J US 47(Suppl): 33–46

Henrickson J (1977) Leaf production and flowering in ocotillos. Cact Succ J US 49: 133–137

Henrickson J (1985) A taxonomic revision of *Chilopsis* (Bignoniaceae). Aliso 11: 179–197

Hoffmann A (1972) Morphology and histology of *Trevoa trinervis* (Rhamnaceae), a drought deciduous shrub from the Chilean matorral. Flora 161: 527–538

Holm T (1894) Adaptation of African plants to climate. Bot Gaz 19: 413–414

Huber B (1935) Xerophyten. Handwörterbuch der Naturwissenschaften 10, 2nd edn. G Fischer, Jena

Hull HM, Shellhorn SJ, Saunier RE (1971) Variations in creosotebush (*Larrea divaricata*) epidermis. J Ariz Acad Sci 6: 196–205

Hull JR, Leonard OA (1964) Physiological aspects of parasitism in mistletoes (*Arceuthobium* and *Phoradendron*). I. The photosynthetic capacity of mistletoe. Plant Physiol 39: 1008–1017

Humphrey RR (1931) Thorn formation in *Fouquieria splendens* and *Idria columnaris.* Bull Torrey Bot Club 58: 263–264

Humphrey RR (1935) A study of *Idria columnaris* and *Fouquieria splendens.* Am J Bot 22: 184–207

Humphrey RR (1974) The boojum and its home. University of Arizona Press, Tucson

Humphrey RR (1975) Phenology of selected Sonoran Desert plants at Punta Cirio, Sonora, Mexico. J Ariz Acad Sci 10: 50–67

Ihlenfeldt H-D (1983) Epidermal structure in Mesembryanthemaceae. Bothalia 14: 931–937

Ihlenfeldt H-D, Hartmann H (1982) Leaf surfaces in Mesembryanthemaceae. In: Cutler DF, Alvin KL, Price CE (eds) The plant cuticle. Academic Press, London, pp 397–423

Iljin WS (1957) Drought resistance in plants and physiological processes. Annu Rev Plant Physiol 8: 257–274

Inouye RS (1980) Density-dependent germination response by seeds of desert annuals. Oecologia 46: 235–238

Inouye RS (1991) Population biology of desert annual plants. In: Polis GA (ed) The ecology of desert communities. University of Arizona Press, Tucson, pp 27–54

Inouye RS, Byers GS, Brown JH (1980) Effects of predation and competition on survivorship, fecundity, and community structure of desert annuals. Ecology 61: 1344–1351

Isebrands JG, Larson PR (1973) Anatomical changes during leaf ontogeny in *Populus deltoides.* Am J Bot 60: 199–208

Ismail AMA (1983) Some factors controlling the water economy of *Zygophyllum quatarense* (Hadidi) growing in Qatar. J Arid Environ 6: 239–246

James LE, Kyhos DW (1961) The nature of the fleshy shoot of *Allenrolfea* and allied genera. Am J Bot 48: 101–108

Jennings DH (1968) Halophytes, succulence and sodium in plants—a unified theory. New Phytol 67: 899–911

Jennings DH (1976) The effects of sodium chloride on higher plants. Biol Rev 51: 453–486

Jessop J (ed) (1981) Flora of Central Australia. Australian Systematic Botany Society, Reed Books, Sydney

Johnson HB (1975) Plant pubescence: an ecological perspective. Bot Rev 41: 233–258

Johnson SC, Brown WV (1973) Grass leaf ultrastructural variations. Am J Bot 60: 727–735

Jones CS (1984) The effect of axis splitting on xylem pressure potentials and water movement in the desert shrub *Ambrosia dumosa* (Gray) Payne (Asteraceae). Bot Gaz 145: 125–131

Jones CS, Lord EM (1982) The development of split axes in *Ambrosia dumosa* (Gray) Payne (Asteraceae). Bot Gaz 143: 446–453

Jones HG (1992) Plants and microclimate, 2nd edn. Cambridge University Press, Cambridge

Jones MM, Turner NC, Osmond CB (1981) Mechanisms of drought resistance. In: Paleg LG, Aspinall D (eds) The physiology and biochemistry of drought resistance in plants. Academic Press, Sydney, pp 15–27

Jordaan A, Kruger H (1992) Leaf surface and anatomy of two xerophytic plants from southern Africa. S Afr J Bot 58: 133–138

Jordan PW, Nobel PS (1981) Seedling establishment of *Ferocactus acanthodes* in relation to drought. Ecology 62: 901–906

Jordan PW, Nobel PS (1984) Thermal and water relations of roots of desert succulents. Ann Bot 54: 705–717

Juhren M, Went FW, Phillips E (1956) Ecology of desert plants. IV. Combined field and laboratory work on germination of annuals in the Joshua Tree National Monument, California. Ecology 37: 318–330

Jurik TW, Chabot JF, Chabot BF (1979) Ontogeny of photosynthetic performance in *Fragaria virginiana* under changing light regimes. Plant Physiol 63: 542–547

Kamerling Z (1914) Welche Pflanzen sollen wir "Xerophyten" nennen? Flora 106: 433–454

Kappen L (1981) Ecological significance of resistance to high temperature. In: Lange OL, Nobel PS, Osmond CB, Ziegler H (eds) Encyclopedia of plant physiology, new ser, vol 12A. Physiological plant ecology I. Responses to the physical environment. Springer, Berlin Heidelberg New York, pp 439–474

Kappen L (1988) Ecophysiological relationships in different climatic regions. In: Galun M (ed) Handbook of lichenology, vol 2. CRC Press, Boca Raton, pp 37–100

Kappen L Rogers RW (1982) Lichens in arid regions. J Hattori Bot Lab 53: 305–307

Kappen L, Lange OL, Schulze E-D, Evenari M, Buschbom U (1972) Extreme water stress and photosynthetic activity of the desert plant *Artemisia herba-alba* Asso. Oecologia 10: 177–182

Kappen L, Oertli JJ, Lange OL, Schulze E-D, Evenari M, Buschbom U (1975) Seasonal and diurnal courses of water relations of the arido-active plant *Hammada scoparia* in the Negev Desert. Oecologia 21: 175–192

Kappen L, Lange OL, Schulze E-D, Evenari M, Buschbom U (1976) Distributional pattern of water relations and net photosynthesis of *Hammada scoparia* (Pomel) Iljin in a desert environment. Oecologia 23: 323–334

Karsten MC (1931) Light receptors in *Mesembryanthemum*. Trans Proc Bot Soc Edinb 30(1): 37–42

Kasapligil B (1961) Foliar xeromorphy of certain geophytic monocotyledons. Madroño 16: 43–70

Kearney TH, Shantz HL (1911) The water economy of dry land crops. US Dep Agric Yearb Agric 10: 331–336

Kee C, Nobel PS (1986) Concomitant changes in high temperature tolerance and heat-shock proteins in desert succulents. Plant Physiol 80: 596–598

Keller B (1925) 'Halophyten- und Xerophyten-Studien'. J Ecol 13: 224–261

Kemp PR, Gardetto PE (1982) Photosynthetic pathway types of evergreen rosette plants (Liliaceae) of the Chihuahuan Desert. Oecologia 55: 149–156

Kessler A (1985) Heat balance climatology. Essenwanger OM (ed) World survey of climatology, general climatology, vol IA. Elsevier Scientific, Amsterdam

Kluge M (1979) The flow of carbon in crassulacean acid metabolism (CAM). In: Gibbs M, Latzko E (eds) Encyclopedia of plant physiology, new ser, vol 6. Photosynthesis II. Springer, Berlin Heidelberg New York, pp 113–125

Kluge M, Ting IP (1978) Crassulacean acid metabolism: analysis of an ecological adaptation. Ecological studies, vol 30. Springer, Berlin Heidelberg New York

Knapp AK, Smith WK (1991) Gas exchange responses to variable sunlight in two Sonoran Desert herbs: comparison with subalpine species. Bot Gaz 152: 269–274

Knoblock IW, Volz PA (1964) Studies in the fern genus *Cheilanthes* Swartz—I. The leaf blade anatomy of some species of the genus. Phytomorphology 14: 508–527
Koller AL, Rost TL (1988a) Leaf anatomy of *Sansevieria* (Agavaceae). Am J Bot 75: 615–633
Koller AL, Rost TL (1988b) Structural analysis of water-storage tissue in leaves of Sansevieria (Agavaceae). Bot Gaz 149: 260–274
Koller D (1956) Germination regulating mechanisms in some desert seeds. III. *Calligonum comosum* L'Her. Ecology 37: 430–433
Koller D (1990) Light-driven leaf movements. Plant Cell Environ 13: 615–632
Körner C, Neumayer M, Menendez-Riedl SP, Smeets-Scheel A (1989) Functional morphology of mountain plants. Flora 182: 353–383
Kramer D (1979) Ultrastructural observations on developing leaf bladder cells of *Mesembryanthemum crystallinum* L. Flora 168: 193–204
Kummerow J (1973) Comparative anatomy of sclerophylls of Mediterranean climatic areas. In: di Castri F, Mooney HA (eds) Mediterranean type ecosystems. Ecological studies, vol 7. Springer, Berlin Heidelberg New York, pp 157–167
Kurtz EB (1958) Survey of some plant waxes of southern Arizona. J Am Oil Chem Soc 35: 465–467
Landsberg JJ, Butler DR (1980) Stomatal response to humidity: implications for transpiration. Plant Cell Environ 3: 29–33
Lange OL (1959) Untersuchungen über Wärmehaushalt und Hitzeresistenz mauretanischer Wüsten- und Savannenpflanzen. Flora 147: 595–651
Lange OL, Zuber M (1977) *Frerea indica*, a stem succulent CAM plant with deciduous C_3 leaves. Oecologia 31: 67–72
Lange OL, Zuber M (1980) Temperaturabhängigkeit des CO_2-Gaswechsels stammsukkulenter Asclepiadaceen mit Säurestoffwechsel. Flora 170: 529–553
Lange OL, Schulze E-D, Kappen L, Buschbom U, Evenari M (1975a) Photosynthesis of desert plants as influenced by internal and external factors. In: Gates DM, Schmerl RB (eds) Perspectives of biophysical ecology. Ecological studies, vol 12. Springer, Berlin Heidelberg New York, pp 121–143
Lange OL, Schulze E-D, Kappen L, Evenari M, Buschbom U (1975b) CO_2 exchange pattern under natural conditions of *Caralluma negevensis*, a CAM plant of the Negev Desert. Photosynthetica 9: 318–326
Lebkuecher JG, Eickmeier WG (1990) Desiccation-induced plant curling and high-irradiance damage in the desert resurrection plant *Selaginella lepidophylla*. Bull Ecol Soc Am 71: 227
Lebkuecher JG, Eickmeier (1991) Reduced photoinhibition with stem curling in the resurrection plant *Selaginella lepidophylla*. Oecologia 88: 597–604
Lee DG (1912) Notes on the anatomy and morphology of *Pachypodium namaquanum* Welw. Ann Bot 26: 929–942
Lekhak HD, Kumar S, Sen DN (1983) Ecology of Indian Desert. IX. An investigation on the behaviour of leaf and stomata in *Capparis decidua* (Forsk.) Edgew. Flora 174: 475–487
Lemée G (1946) Recherches sur l'économie d'eau chez quelques les sous-arbrisseaux xéromorphes de lands. Ann Sci Nat Bot 11(7): 53–85
Lemée G (1952) Sur l'efficacité de l'enroulement des feuilles de Graminées contre la transpiration. Compt Rend Acad Sci Paris 230: 1201–1203
Lemée G (1954) L'économie de l'eau chez quelques graminées vivaces du Sahara septentrional. Vegetatio 5–6: 534–541
Levitt J (1972) Responses of plants to environmental stresses. Academic Press, New York
Levitt J (1980) Responses of plants to environmental stresses, 2nd edn. Academic Press, New York
Lewis DA, Nobel PS (1977) Thermal energy exchange model and water loss of a barrel cactus, *Ferocactus acanthodes*. Plant Physiol 60: 609–616
Liphschitz N, Waisel Y (1970a) Environmental effects on wood production and cambial activity in *Ziziphus spina-christi* (L.) Willd. Isr J Bot 19: 592–598
Liphschitz N, Waisel Y (1970b) Effects of environment on relations between extension and cambial growth of *Populus euphratica* Oliv. New Phytol 69: 1059–1064
Liphschitz N, Waisel Y (1974) Existence of salt glands in various genera of the Gramineae. New Phytol 73: 507–513
Liphschitz N, Waisel Y (1982) Adaptation of plants to saline excretion and glandular structure. In: Sen DN, Rajpurohit KS (eds) Contributions to the ecology of halophytes. Dr W Junk, The Hague, pp 197–214

Loik ME, Nobel PS (1991) Water relations and mucopolysaccharide increases for a winter hardy cactus during acclimation to subzero temperatures. Oecologia 88: 340–346

Longstreth DJ, Hartsock TL, Nobel PS (1980) Mesophyll cell properties for some C_3 and C_4 species with high photosynthetic rates. Physiol Plant 48: 494–498

Lopez FB, Nobel PS (1991) Root hydraulic conductivity of two species in relation to root age, temperature, and soil water status. J Exp Bot 42: 143–149

Lorimer GH (1981) The carboxylation and oxygenation of ribulose 1,5-bisphosphate: the primary events in photosynthesis and photorespiration. Annu Rev Plant Physiol 32: 349–383

Lösch R, Tenhunen JD (1981) Stomatal responses to humidity—phenomenon and mechanism. In: Jarvis PG, Mansfield TA (eds) Stomatal physiology. Cambridge University Press, Cambridge

Ludwig JA (1987) Primary productivity in arid lands: myths and realities. J Arid Environ 13: 1–7

Lüttge U, Smith JAC (1984) Structural, biophysical, and biochemical aspects of the role of leaves in plant adaptation to salinity and water stress. In: Staples RC, Toenniessen GH (eds) Salinity tolerance in plants. Wiley-Interscience, New York, pp 125–150

Lyshede OB (1977a) Anatomical features of some stem assimilating desert plants of Israel. Bot Tidsskr 71: 225–230

Lyshede OB (1977b) Structure of the epidermal and subepidermal cells off some desert plants of Israel. *Anabasis articulata* and *Calligonum comosum.* Isr J Bot 26: 1–10

Lyshede OB (1977c) Structure and function of trichomes in *Spartocytisus filipes.* Bot Not 129: 395–404

Lyshede OB (1977d) Studies on the mucilaginous cells in the leaf of *Spartocytisus filipes* W.B. Planta 133: 255–260

Lyshede OB (1978) Studies on outer epidermal cell walls with microchannels in a xerophytic species. New Phytol 80: 421–426

Lyshede OB (1979) Xeromorphic features of three stem assimilants in relation to their ecology. J Linn Soc Bot 78: 85–99

Mabry TJ, Difeo DR Jr, Sakakibara M, Bohnstedt CF Jr, Seigler D (1977) The natural products chemistry of *Larrea.* In: Mabry TJ, Hunziker JH, Difeo DR Jr (eds) Creosote bush: biology and chemistry of *Larrea* in New World deserts. US/IBP synthesis ser 6. Dowden, Hutchinson, and Ross, Stroudsburg, pp 115–134

MacDougal DT, Spalding ES (1910) The water-balance of succulent plants. Carnegie Inst Wash Pub 141, Washington, DC

Manning SJ, Barbour MG (1988) Root systems, spatial patterns, and competition for soil moisture between two desert subshrubs. Am J Bot 75: 885–893

Marloth R (1909) Die Schutzmittel der Pflanzen gegen übermässige Insolation. Ber Dtsch Bot Ges 27: 362–371

Marshall JD, Ehleringer JR (1990) Are xylem-tapping mistletoes partially heterotrophic? Oecologia 84: 244–248

Martin CE, Peters EA (1984) Functional stomata of the atmospheric epiphyte *Tillandsia usneoides* L. Bot Gaz 145: 502–507

Martin CE, Siedow JN (1981) Crassulacean acid metabolism in the epiphyte *Tillandsia usneoides* L. (Spanish moss). Response of CO_2 exchange to controlled environmental conditions. Plant Physiol 68: 335–339

Martin JT, Juniper BE (1970) The cuticles of plants. Edward Arnold, London

Maury P (1887) Anatomie comparée de quelques espèces caractéristiques du Sahara algérien. Assoc Franç Avancem Sci C16e Sess Toulouse 2: 604–632

Mauseth JD (1990) Morphogenesis in a highly reduced plant: the endophyte of *Tristerix aphyllus* (Loranthaceae). Bot Gaz 151: 348–353

Mauseth JD, Montenegro G, Walckowiak AM (1984) Studies of the holoparasitic *Tristerix aphyllus* (Loranthaceae) infecting *Trichocereus chilensis* (Cactaceae). Can J Bot 62: 847–857

Maximov NA (1929) The plant in relation to water. A study of the physiological basis of drought resistance. Allen and Unwin, London

Maximov NA (1931) The physiological significance of the xeromorphic structure of plants. J Ecol 19: 273–282

McCleary JA (1968) The biology of desert plants. In: Brown GW Jr (ed) Desert biology, vol 1. Academic Press, New York, pp 141–195

McGinnies WG (1979) Arid-land ecosystems—common features throughout the world. In: Perry RA, Goodall DW (eds) Arid-land ecosystems: structure, functioning and management, vol 1. International Biological Programme 16. Cambridge University Press, Cambridge
McGinnies WG, Goldman BJ, Paylore P (eds) (1968) Deserts of the world: an appraisal of research into their physical and biological environments. University of Arizona Press, Tucson
Meinzer FC, Wisdom CS, Gonzalez-Coloma A, Rundel PW, Shultz LM (1990) Effects of leaf resin on stomatal behaviour and gas exchange of *Larrea tridentata* (DC.) Cov. Funct Ecol 4: 579–584
Merino J, Field C, Mooney HA (1984) Construction and maintenance costs of Mediterranean-climate evergreen and deciduous leaves. II. Biochemical pathway analysis. Ecol Plant 5: 211–229
Metcalfe CR, Chalk L (1950) Anatomy of the dicotyledons, 2 vols. Clarendon Press, Oxford
Metcalfe CR, Chalk L (1989) Anatomy of the docotyledons, vol 2, 2nd edn. Clarendon Press, Oxford
Meyer SE (1986) The ecology of gypsophile endemism in the eastern Mojave Desert. Ecology 67: 1303–1313
Michell MR (1912) On the comparative anatomy of the genera *Ceraria* and *Portulacaria*. Ann Bot 26: 1111–1122
Monson RK, Smith SD (1982) Seasonal water potential components of Sonoran Desert plants. Ecology 63: 113–123
Mooney HA (1980a) Photosynthetic plasticity of populations of *Heliotropium curassavicum* L. originating from differing thermal regimes. Oecologia 45: 372–376
Mooney HA (1980b) Seasonality and gradients in the study of stress adaptations. In: Turner NC, Kramer PJ (eds) Adaptations of plants to water and high temperature stress. Wiley-Interscience, New York, pp 279–294
Mooney HA (1981) Primary production in mediterranean-climate regions. In: di Castri F, Goodall DW, Specht RL (eds) Mediterranean-type shrublands. Elsevier, Amsterdam, pp 249–255
Mooney HA (1983) Carbon-gaining capacity and allocation patterns of Mediterranean-climate plants. In: Kruger FJ, Mitchell DT, Jarvis JUM (eds) Mediterranean-type ecosystem: the role of nutrients. Springer, Berlin Heidelberg New York, pp 103–119
Mooney HA (1989) Chaparral physiological ecology—paradigms revisited. In: Keeley JE (ed) The California chaparral: paradigms reexamined. Natural History Museum of Los Angeles County Science Series 34, Los Angeles, pp 85–90
Mooney HA, Ehleringer JR (1978) The carbon gain benefits of solar tracking in a desert annual. Plant Cell Environ 1: 307–311
Mooney HA, Gulmon SL (1979) Environmental and evolutionary constraints on the photosynthetic characteristics of higher plants. In: Solbrig OT, Jain S, Johnson HB, Raven PH (eds) Topics in plant population biology. Columbia University Press, New York, pp 316–337
Mooney HA, Gulmon SL (1982) Constraints on leaf structure and function in reference to herbivory. BioScience 32: 198–206
Mooney HA, Strain BR (1964) Bark photosynthesis in ocotillo. Madroño 17: 230–233
Mooney HA, Björkman O, Berry J (1975) Photosynthetic adaptations to high temperatures. In: Hadley NF (ed) Environmental physiology of desert organisms. Dowden, Hutchinson and Ross, Stroudsburg, pp 138–150
Mooney HA, Björkman O, Ehleringer J, Berry JA (1976a) Photosynthetic capacity of *in situ* Death Valley plants. Carnegie Inst Wash Yearb 1975: 410–413
Mooney HA, Ehleringer J, Berry J (1976b) High photosynthetic capacity of a winter annual in Death Valley. Science 194: 322–324
Mooney HA, Ehleringer JR, Björkman O (1977a) The energy balance of leaves of the evergreen desert shrub *Atriplex hymenelytra*. Oecologia 29: 301–310
Mooney HA, Simpson BB, Solbrig OT (1977b) Phenology, morphology, physiology. In: Simpson BB (ed) Mesquite: its biology in two desert scrub ecosystems. Dowden, Hutchinson and Ross, Stroudsburg, pp 26–43
Mooney HA, Weisser PJ, Gulmon SL (1977c) Environmental adaptations of the Atacaman Desert cactus *Copiapoa haseltoniana*. Flora 166: 117–124
Mooney HA, Björkman O, Collatz GJ (1978) Photosynthetic acclimation to temperature in the desert shrub *Larrea divaricata*. I. Carbon dioxide exchange characteristics of intact leaves. Plant Physiol 61: 406–410
Mooney HA, Gulmon SL, Ehleringer J, Rundel PW (1980a) Atmospheric water uptake by an Atacaman Desert shrub. Science 209: 693–694

Mooney HA, Gulmon SL, Rundel PW, Ehleringer J (1980b) Further observations on the water relations of *Prosopis tamarugo* of the northern Atacama Desert. Oecologia 44: 177–180

Mooney HA, Field C, Gulmon SL, Bazzaz FA (1981) Photosynthetic capacity in relation to leaf position in desert versus old-field annuals. Oecologia 50: 109–112

Morello J (1955) Estudios botánicos en las regiones áridas de la Argentina. I. Ambiente, morfología y anatomía de cuatro arbustos resinosos de follaje permanente del Monte. Rev Agron Noroeste Argent 1: 301–370

Mott KA, Michaelson O (1991) Amphistomaty as an adaptation to high light intensity in *Ambrosia cordifolia* (Compositae). Am J Bot 78: 76–79

Mott KA, O'Leary JW (1984) Stomatal behavior and CO_2 exchange characteristics in amphistomatous leaves. Plant Physiol 74: 47–51

Mott KA, Gibson AC, O'Leary JW (1982) The adaptive significance of amphistomatic leaves. Plant Cell Environ 5: 455–460

Mozingo HN, Comanor PL (1975) Implications of the thermal response of *Ferocactus acanthodes*. In: Glass C, Foster R (eds) 1975 Yearbook. Cact Succ J US 47(Suppl): 22–28

Mulroy TW (1979) Spectral properties of heavily glaucous and non-glaucous leaves of a succulent rosette-plant. Oecologia 38: 349–357

Mulroy TW, Rundel PW (1977) Annual plants: adaptations to desert environments. BioScience 27: 109–114

Munns R, Greenway H, Kirst GO (1983) Halotolerant eukaryotes. In: Lange OL, Nobel PS, Osmond CB, Ziegler H (eds) Encyclopedia of plant physiology, new ser, vol 12C. Physiological plant ecology III. Responses to the chemical and biological environment. Springer, Berlin Heidelberg New York, pp 59–135

Musick HB (1975) Barrenness of desert pavement in Yuma County, Arizona. J Ariz Acad Sci 10: 24–28

Napp-Zinn K (1966) Anatomie des Blattes. Blattanatomie der Gymnospermen, Bd 8, Teil 1. Handbuch der Pflanzenanatomie. Borntraeger, Berlin

Napp-Zinn K (1973) Anatomie des Blattes. Blattanatomie der Angiospermen, Bd 8, Teil 2A(1). Handbuch der Pflanzenanatomie. Borntraeger, Berlin

Napp-Zinn K (1974) Anatomie des Blattes. Blattanatomie der Angiospermen, Bd 8, Teil 2A(2). Handbuch der Pflanzenanatomie. Borntraeger, Berlin

Nash III TH, White SL, Marsh JE (1977) Lichen and moss distribution and biomass in hot desert ecosystems. Bryologist 80: 470–475

Nedoff JA, Ting IP, Lord LM (1985) Structure and function of the green stem tissue in ocotillo (*Fouquieria splendens*). Am J Bot 72: 143–151

Neufeld HS, Meinzer FC, Wisdom CS, Sharifi MR, Rundel PW, Neufeld MS, Goldring Y, Cunningham GL (1988) Canopy architecture of *Larrea tridentata* (DC.) Cov., a desert shrub: foliage orientation and direct beam radiation interception. Oecologia 75: 54–60

Nicholson SE (1978) Climatic variations in the Sahel and other African regions during the past five centuries. J Arid Environ 1: 3–24

Niklas KJ (1992) Plant biomechanics. University of Chicago Press, Chicago

Nilsen ET (1995) Stem photosynthesis: extent, pattern, and role in plant carbon economy. In: Gartner B (ed) Plant stems, physiology and functional morphology. Academic Press, New York, pp 223–240

Nilsen ET, Sharifi MR (1994) Seasonal acclimation of stem photosynthesis in woody legume species from the Mojave and Sonoran deserts of California. Plant Physiol 105: 1385–1391

Nilsen ET, Rundel PW, Sharifi MR (1981) Summer water relations of the desert phreatophyte *Prosopis glandulosa* in the Sonoran Desert of California. Oecologia 50: 271–276

Nilsen ET, Sharifi MR, Rundel PW, Jarrell WM, Virginia RA (1983) Diurnal and seasonal water relations of the desert phreatophyte *Prosopis glandulosa* (honey mesquite) in the Sonoran Desert of California. Ecology 64: 1381–1393

Nilsen ET, Sharifi MR, Rundel PW (1984) Comparative water relations of phreatophytes in the Sonoran Desert of California. Ecology 65: 767–778

Nilsen ET, Sharifi MR, Rundel PW, Virginia RA (1986) Influences of microclimatic conditions and water relations on seasonal leaf dimorphism of *Prosopis glandulosa* var. *torreyana* in the Sonoran Desert, California. Oecologia 69: 95–100

Nilsen ET, Meinzer FC, Rundel PW (1989) Photosynthetic response to light, temperature, humidity, and CO_2 in the stem photosynthetic tree *Psorothamnus spinosus*. Oecologia 79: 480–486

Nilsen ET, Sharifi MR, Rundel PW, Forseth IN, Ehleringer JR (1990) Water relations of stem succulent trees in north-central Baja California. Oecologia 82: 299–303

Nilsen ET, Sharifi MR, Rundel PW (1991) Quantitative phenology of warm desert legumes: seasonal growth of six *Prosopis* species at the same site. J Arid Environ 20: 299–311

Nobel PS (1974a) Boundary layers of air adjacent to cylinders. Plant Physiol 54: 171–181

Nobel PS (1974b) Introduction to biophysical plant ecology. WH Freeman, San Francisco

Nobel PS (1976a) Photosynthetic rates of sun *versus* shade leaves of *Hyptis emoryi* Torr. Plant Physiol 58: 218–223

Nobel PS (1976b) Water relations and photosynthesis of a desert CAM plant, *Agave deserti*. Plant Physiol 58: 576–582

Nobel PS (1978a) Microhabitat, water relations and photosynthesis of a desert fern *Notholaena parryi*. Oecologia 31: 293–309

Nobel PS (1978b) Surface temperatures of cacti—influences of environmental and morphological factors. Ecology 59: 986–995

Nobel PS (1980a) Interception of photosynthetically active radiation by cacti of different morphology. Oecologia 45: 160–166

Nobel PS (1980b) Morphology, nurse plants, and minimum apical temperatures for young *Carnegiea gigantea*. Bot Gaz 141: 188–191

Nobel PS (1980c) Morphology, surface temperatures, and northern limit of columnar cacti in the Sonoran Desert. Ecology 61: 1–7

Nobel PS (1980d) Water vapor conductance and CO_2 uptake for leaves of a C_4 desert grass, *Hilaria rigida*. Ecology 61: 252–258

Nobel PS (1981a) Influences of photosynthetically active radiation on cladode orientation, stem tilting, and height of cacti. Ecology 62: 982–990

Nobel PS (1981b) Spacing and transpiration of various sized clumps of a desert grass, *Hilaria rigida*. J Ecol 69: 735–742

Nobel PS (1982a) Interaction between morphology, PAR interception, and nocturnal acid accumulation in cacti. In: Ting IP, Gibbs M (eds) Crassulacean acid metabolism. American Society of Plant Physiologists, Rockville, pp 260–277

Nobel PS (1982b) Orientation of terminal cladodes of platyopuntias. Bot Gaz 143: 219–224

Nobel PS (1983a) Biophysical plant physiology and ecology. WH Freeman, San Francisco

Nobel PS (1983b) Low and high temperature influences on cacti. In: Marcelle R, Clijsters H, van Poucke M (eds) Effects of stress on photosynthesis. Nijhoff/Dr W Junk, The Hague, pp 165–174

Nobel PS (1983c) Spines influences on PAR interception, stem temperature, and nocturnal acid accumulation by cacti. Plant Cell Environ 6: 153–159

Nobel PS (1984) Extreme temperatures and thermal tolerances for seedlings of desert succulents. Oecologia 62: 310–317

Nobel PS (1985) Desert succulents. In: Chabot BF, Mooney HA (eds) Physiological ecology of North American plant communities. Chapman and Hall, New York, pp 181–197

Nobel PS (1986) Form and orientation in relation to PAR interception by cacti and agaves. In: Givnish TJ (ed) On the economy of plant form and function. Cambridge University Press, Cambridge, pp 83–103

Nobel PS (1988) Environmental biology of agaves and cacti. Cambridge University Press, New York

Nobel PS (1989a) Shoot temperatures and thermal tolerances for succulent species of *Haworthia* and *Lithops*. Plant Cell Environ 12: 643–651

Nobel PS (1989b) Temperature, water availability, and nutrient levels at various soil depths—consequences for shallow-rooted desert succulents, including nurse plant effects. Am J Bot 76: 1486–1492

Nobel PS (1991a) Physicochemical and environmental plant physiology. Academic Press, San Diego

Nobel PS (1991b) Achievable productivities of CAM plants; basis for high values compared with C_3 and C_4. New Phytol 119: 183–205

Nobel PS (1991c) Ecophysiology of roots of desert plants, with special emphasis on agaves and cacti. In: Waisel Y, Eshel A, Kafkafi U (eds) Plant roots: the hidden half. Marcel Dekker, New York, pp 839–866

Nobel PS (1994a) Remarkable agaves and cacti. Oxford University Press, New York

Nobel PS (1994b) Root-soil responses to water pulses in dry environments. In: Caldwell MM, Pearcy RW (eds) Exploitation of environmental heterogeneity by plants: ecophysiological processes above- and belowground. Academic Press, San Diego, pp 285–304

Nobel PS, Cui M (1992) Hydraulic conductances of the soil, the root-soil air gap, and the root: changes for desert succulents in drying soil. J Exp Bot 43: 319–326

Nobel PS, Franco AC (1986) Effect of nurse plants on the microhabitat and growth of cacti. J Ecol 77: 870–886

Nobel PS, Huang B (1992) Hydraulic and structural changes for lateral roots of two desert succulents in response to soil drying and rewetting. Int J Plant Sci 153(3): S163–170

Nobel PS, Longstreth DJ (1981) Effects of environmental factors on leaf anatomy, mesophyll cell conductance, and photosynthesis. In: Akoyunoglou G (ed) Photosynthesis. VI. Photosynthesis and productivity, photosynthesis and environment. Balaban International Science Service, Philadelphia, pp 245–254

Nobel PS, Sanderson J (1984) Rectifier-like activities of roots of two desert succulents. J Exp Bot 35: 727–737

Nobel PS, Smith SD (1983) High and low temperature tolerances and their relationships to distribution of agaves. Plant Cell Environ 6: 711–719

Nobel PS, Walker DB (1985) Structure of leaf photosynthetic tissue. In: Barber J, Baker NR (eds) Photosynthetic mechanisms and the environment. Elsevier, Amsterdam, pp 501–536

Nobel PS, Zaragoza LJ, Smith WK (1975) Relation between mesophyll surface area, photosynthetic rate, and illumination level during development for leaves of *Plectranthus parviflorus* Henckel. Plant Physiol 55: 1067–1070

Nobel PS, Geller GN, Kee SC, Zimmerman AD (1986) Temperatures and thermal tolerances for cacti exposed to high temperatures near the soil surface. Plant Cell Environ 9: 279–287

Nobel PS, Schulte PJ, North GB (1990) Water influx characteristics and hydraulic conductivity for roots of *Agave deserti* Engelm. J Exp Bot 41: 409–415

Nobel PS, Cavelier J, Andrade JL (1992a) Mucilage in cacti: its apoplastic capacitance, associated solutes, and influences on tissue water relations. J Exp Bot 43: 641–648

Nobel PS, Miller PM, Graham EA (1992b) Influence of rocks on soil temperature, soil water potential, and rooting patterns for desert succulents. Oecologia 92: 90–96

Nobel PS, Forseth IN, Long SP (1993) Canopy structure and light interception. In: Hall DO, Scurlock JMO, Bolhar-Nordenkampt HR, Leegood RC, Long SP (eds) Photosynthesis and production in a changing environment: a field and laboratory manual. Chapman and Hall, London, pp 79–90

North GB (1992) Water uptake and storage in diverse arid-land plants. PhD Dissertation, University of California, Los Angeles

North GB, Nobel PS (1992) Drought-induced changes in hydraulic conductivity and structure in roots of *Ferocactus acanthodes* and *Opuntia ficus-indica*. New Phytol 120: 9–19

Oberlander TM (1979) Characterization of arid climates according to combined water balance parameters. J Arid Environ 2: 219–241

Odening WR, Strain BR, Oechel WC (1974) The effect of decreasing water potential on net CO_2 exchange of intact desert shrubs. Ecology 55: 1086–1095

Oertli JJ (1986) The effect of cell size on cell collapse under negative turgor pressure. J Plant Physiol 124: 365–370

Ogren WL (1984) Photorespiration: pathways, regulation and modification. Annu Rev Plant Physiol 35: 415–444

Oppenheimer HR (1960) Adaptation to drought: xerophytism. Plant-water relationships in arid and semi-arid conditions. Arid Zone Res (UNESCO) 15: 105–138

Orians GH, Solbrig OT (1977a) A cost-income model of leaves and roots with special reference to arid and semiarid areas. Am Nat 111: 677–690

Orians GH, Solbrig OT (eds) (1977b) Convergent evolution in warm deserts. Dowden, Hutchinson and Ross, Stroudsburg

Oross JW, Leonard RT, Thomson WW (1985) Flux rate and a secretion model for salt glands of grasses. Isr J Bot 34: 66–77

Orshan G (1954) Surface reduction and it significance as a hydroecological factor. J Ecol 42: 442–444

Orshan G (1963) Seasonal dimorphism of desert and mediterranean chamaephytes and its significance as a factor of their economy. In: Ruther AJ, Whitehead FW (eds) Water relations of plants. Blackwell, Oxford, pp 206–222

Osmond CB (1974) Leaf anatomy of Australian saltbushes in relation to photosynthetic pathways. Aust J Bot 22: 39–44
Osmond CB (1978) Crassulacean acid metabolism: a curiosity in context. Annu Rev Plant Physiol 29: 379–414
Osmond CB, Lüttge U, West U, Pallaghy CK, Shacher-Hill B (1969) Ion absorption in *Atriplex* leaf tissue: II, secretion of ions to epidermal bladders. Aust J Biol Sci 22: 797–814
Osmond CB, Björkman O, Anderson DJ (1980) Physiological processes in plant ecology. Toward a synthesis with *Atriplex*. Springer, Berlin Heidelberg New York
Osmond CB, Winter K, Ziegler H (1982) Functional significance of different pathways of CO_2 fixation in photosynthesis. In: Lange OL, Nobel PS, Osmond CB, Ziegler H (eds) Encyclopedia of plant physiology, new ser, vol 12B. Physiological plant ecology II. Water relations and carbon assimilation. Springer, Berlin Heidelberg New York, pp 479–547
Osmond CB, Smith SD, Gui-Ying B, Starkey TD (1987) Stem photosynthesis in a desert ephemeral, *Eriogonum inflatum*. Characterization of leaf and stem CO_2 fixation and H_2O vapor exchange under controlled conditions. Oecologia 72: 542–549
Pallaghy CK (1970) Salt relations of *Atriplex* leaves. In: Jones R (ed) The biology of *Atriplex*. Division of Plant Industry, Commonwealth Scientific and Industrial Research Organization, Canberra, pp 57–62
Parkhurst DF (1978) The adaptive significance of stomatal occurrence on one or both surfaces of leaves. J Ecol 66: 367–383
Pavlik BM (1980) Patterns of water potential and photosynthesis of desert sand dune plants, Eureka Valley, California. Oecologia 46: 147–156
Payne WW (1970) Helicocytic and allelocytic stomata: unrecognized patterns in the Dicotyledonae. Am J Bot 57: 140–147
Pearcy RW, Harrison AT (1974) Comparative photosynthetic and respiratory gas exchange characteristics of *Atriplex lentiformis* (Torr.) Wats. in coastal and desert habitats. Ecology 55: 1104–1111
Pearcy RW, Harrison AT, Mooney HA, Björkman O (1974) Seasonal changes in net photosynthesis of *Atriplex hymenelytra* shrubs growing in Death Valley, California. Oecologia 17: 111–121
Pearcy RW, Berry JA, Fork DC (1977) Effect of growth temperature on the thermal stability of the photosynthetic apparatus of *Atriplex lentiformis* (Torr.) Wats. Plant Physiol 59: 873–878
Pearcy RW, Tumosa N, Williams K (1981) Relationships between growth, photosynthesis and competitive interactions for a C_3 and C_4 plant. Oecologia 48: 371–376
Pearson LC, Lawrence DB (1958) Photosynthesis in aspen bark. Am J Bot 45: 383–387
Pillai A, Pillai SK (1977) Some aspects of the anatomy of *Salvadora oleoides* Dcne. Flora 166: 211–218
Plowes DCH (1990) An introduction to stapeliad genera. Cact Succ J US 62: 111–129
Poljakoff-Mayber A, Gale J (eds) (1975) Plants in saline environments. Ecological studies, vol 15. Springer, Berlin Heidelberg New York
Porembski S, Martens-Aly B, Barthlott W (1991) Surface/volume-ratios of plants with special consideration of succulents. Beitr Biol Pflanz 66: 189–209
Prendergast HDV (1989) Geographical distribution of C_4 acid decarboxylation types and associated structural variants in native Australian C_4 grasses (Poaceae). Aust J Bot 37: 253–273
Prendergast HDV, Hattersley PW (1987) Australian C_4 grasses (Poaceae): leaf blade anatomical features in relation to C_4 acid decarboxylation types. Aust J Bot 35: 355–382
Price PW (1982) Wild buckwheat, *Eriogonum inflatum* (Polygonaceae): an enigmatic plant. Southwest Nat 27: 247–253
Price RS (1911) The roots of some North African desert grasses. New Phytol 10: 328–339
Purer EA (1936) Studies of certain coastal sand dune plants of southern California. Ecol Monogr 6: 1–88
Pyykko M (1966) The leaf anatomy of East Patagonian xeromorphic plants. Ann Bot Fenn 3: 453–622
Rauh W (1961) Weitere Untersuchungen an Didiereaceen. Teil 1: Beitrag zur Kenntnis der Wuchsformen der Didiereaceen, unter besonderer Berücksichtigung neuer Arten. Heidelb Akad Wiss Math-Nat Kl 7: 1–118, 185–300
Rauh W (1974) Window-leaved succulents. Cact Succ J US 46: 12–25

Rauh W (1985) The Peruvian-Chilean deserts. In: Evenari M, Noy-Meir I, Goodall DW (eds) Hot deserts and arid shrublands. Ecosystems of the world, vol 12A. Elsevier, The Hague, pp 239–260
Rauh W, Hutchison PC (1973) *Peperomia columella*, a new, dwarf species from northern Peru. Cact Succ J US 45: 152–156
Raunkiaer C (1934) The life forms of plants and statistical plant geography. Clarendon Press, Oxford
Redmann RE (1985) Adaptation of grasses to water stress: leaf rolling and stomate distribution. Ann MO Bot Gard 72: 833–842
Reimold RJ, Queen WH (eds) (1974) Ecology of halophytes. Academic Press, New York
Richardson PE (1970) Morphology of the Crossosomataceae. I. Leaf, stem and node. Bull Torrey Bot Club 97: 34–39
Riveros F de, Hoffmann A, Avila G, Aljaro ME, Araya S, Hoffmann AE, Montenegro G (1976) Comparative morphological and ecophysiological aspects of two sclerophyllous Chilean shrubs. Flora 165: 223–234
Robberecht R, Caldwell MM (1978) Leaf epidermal transmittance of ultraviolet radiation and its implications for plant sensitivity to ultraviolet-radiation induced injury. Oecologia 32: 277–287
Robinson MD, Seely MK (1980) Physical and biotic environments of the southern Namib dune ecosystem. J Arid Environ 3: 183–203
Robinson WJ (1904) The spines of *Fouquieria*. Bull Torrey Bot Club 31: 45–50
Rodin RJ (1958a) Leaf anatomy of *Welwitschia*. I. Early development of the leaf. Am J Bot 45: 90–95
Rodin RJ (1958b) Leaf anatomy of *Welwitschia*. II. A study of mature leaves. Am J Bot 45: 96–103
Rogers RW (1977) Lichens of hot arid and semi-arid lands. In: Seaward MRD (ed) Lichen ecology. Academic Press, New York, pp 211–252
Rouschal E (1938) Eine physiologische Studie an *Cetarch officinarum* Willd. Flora 132: 305–318
Rowley GD (1987) Caudiciform and pachycaul succulents. Strawberry Press, Mill Valley
Rundel PW (1978) Ecological relationships of desert fog zone lichens. Bryologist 81: 277–293
Rundel PW (1982) Water uptake by organs other than roots. In: Lange OL, Nobel PS, Osmond CB, Ziegler H (eds) Encyclopedia of plant physiology, new ser, vol 12B. Physiological plant ecology II. Water relations and carbon assimilation. Springer, Berlin Heidelberg New York, pp 111–134
Rundel PW (1988) Water relations. In: Galun M (ed) Handbook of lichenology, vol 2. CRC Press, Boca Raton, pp 17–36
Rundel PW, Franklin T (1991) Vines in arid and semi-arid ecosystems. In: Putz FE, Mooney HA (eds) The biology of vines. Cambridge University Press, Cambridge, pp 337–356
Rundel PW, Gibson AC (1996) Ecological communities and processes in a Mojave Desert ecosystem. Cambridge University Press, Cambridge
Rundel PW, Lange OL (1980) Water relations and photosynthetic response of a desert moss. Flora 169: 329–335
Rundel PW, Nobel PS (1991) Structure and function in desert roots. In: Atkinson D (ed) Plant root growth: an ecological perspective. Spec Pub 10, British Ecological Society. Blackwell, Oxford, pp 349–378
Rundel PW, Stichler W, Zander RH, Ziegler H (1979) Carbon and hydrogen isotope ratios of bryophytes from arid humid regions. Oecologia 44: 91–94
Rundel PW, Ehleringer J, Mooney HA, Gulmon SL (1980) Patterns of drought response in leaf-succulent shrubs of the coastal Atacama Desert in northern Chile. Oecologia 46: 196-200
Rundel PW, Dillon MO, Palma B, Mooney HA, Gulmon SL, Ehleringer JR (1991) The phytogeography and ecology of the coastal Atacama and Peruvian deserts. Aliso 13: 1–49
Runyon EH (1934) The organization of the creosote bush with respect to drought. Ecology 15: 128–138
Rutherford RJ (1970) The anatomy and cytology of *Pilostyles thurberi* Gray (Rafflesiaceae). Aliso 7: 263–288
SaadEddin R Doddema H (1986) Anatomy of the "extreme" halophyte *Arthrocnemum fruticosum* (L.) Moq. in relation to its physiology. Ann Bot 57: 531–544
Sabnis TS (1919–1921) The physiological anatomy of the plants of the Indian Desert. J Indian Bot Soc 1: 33–48, 65–83, 97–113, 183–211, 237–251, 277–295; 2: 1–20, 61–79, 93–115, 157–173, 217–235, 271–299

Salema R (1967) On the occurrence of periderm in the leaves of *Welwitschia mirabilis.* Can J Bot 45: 1469–1471

Sankhla N, Ziegler H, Vyas OP, Stichler W, Trimborn P (1975) Eco-physiological studies on Indian arid zone plants V. A screening of some species for the C_4-pathway of photosynthetic CO_2-fixation. Oecologia 21: 123–129

Schanderl H (1935) Untersuchungen über die Lichtverhältnisse im Innern von Hartlaub- und Sukkulentenblättern. Planta 24: 454–469

Schery RW (1972) Plants for man, 2nd edn. Prentice Hall, Englewood Cliffs

Schirmer U, Breckle S-W (1982) The role of bladders for salt removal in some Chenopodiaceae (mainly *Atriplex* species). In: Sen DN, Rajpurohit KS (eds) Contributions to the ecology of halophytes. Dr W Junk, The Hague, pp 215–231

Schmidt J, Kaiser WM (1987) Response of the succulent leaves of *Peperomia magnoliaefolia* to dehydration. Plant Physiol 83: 190–194

Schmitt AK, Martin CE, Loeschen VS, Schmitt A (1993) Mid-summer gas exchange and water relations of seven C_3 species in a desert wash in Baja California, Mexico. J Arid Environ 24: 155–164

Schmucker T (1931) Zur Ökologie der Fensterblätter. Planta 13: 1–17

Schouw F (1823) Grundzüge einer allgemeinen Pflanzengeographie. Berlin

Schulte PJ, Nobel PS (1989) Responses of a CAM plant to drought and rainfall: capacitance and osmotic pressure influences on water movement. J Exp Bot 40: 61–70

Schulte PJ, Gibson AC, Nobel PS (1989a) Water flow in vessels with simple or compound perforation plates. Ann Bot 64: 171–178

Schulte PJ, Smith JAC, Nobel PS (1989b) Water storage and osmotic pressure influence on the water relations of a dicotyledonous desert succulent. Plant Cell Environ 12: 831–842

Schulze E-D (1986) Carbon dioxide and water vapor exchange in response to drought in the atmosphere and in the soil. Annu Rev Plant Physiol 37: 247–274

Schulze E-D, Ehleringer JR (1984) The effects of nitrogen supply on the growth and water-use efficiency of xylem-tapping mistletoes. Planta 162: 268–275

Schulze E-D, Schulze I (1976) Distribution and control of photosynthetic pathways in plants growing in the Namib Desert, with special regard to Welwitschia mirabilis Hook. fil. Madoqua (Windhoek) 9(3): 5–13

Schulze E-D, Lange OL, Evenari M, Kappen L, Buschbom U (1974) The role of air humidity and leaf temperature in controlling stomatal resistance of *Prunus armeniaca* L. under desert conditions. I. A simulation of the daily course of stomatal resistance. Oecologia 17: 159–170

Schulze E-D, Lange OL, Kappen L, Evenari M, Buschbom U (1975a) The role of air humidity and leaf temperature in controlling stomatal resistance of *Prunus armeniaca* L. under desert conditions. II. The significance of leaf water status and internal carbon dioxide concentration. Oecologia 18: 219–233

Schulze E-D, Lange OL, Evenari M, Kappen L, Buschbom U (1975b) The role of air humidity and leaf temperature in controlling stomatal resistance of *Prunus armeniaca* L. under desert conditions. III. The effect on water use efficiency. Oecologia 19: 303–314

Schulze E-D, Lange OL, Evenari M, Kappen L, Buschbom U (1976a) An empirical model of net photosynthesis for the desert plant *Hammada scoparia* (Pomel) Iljin. I. Description and test of the model. Oecologia 22: 355–372

Schulze E-D, Ziegler H, Stichler W (1976b) Environmental control of crassulacean acid metabolism in *Welwitschia mirabilis* Hook. fil. in its range of natural distribution in the Namib Desert. Oecologia 24: 323–334

Schulze E-D, Eller BM, Thomas DA, von Willert DJ, Brinckmann E (1980) Leaf temperatures and energy balance of *Welwitschia mirabilis* in its natural habitat. Oecologia 44: 258–262

Schulze E-D, Lange OL, Ziegler H (1991) Carbon and nitrogen isotope ratios of mistletoes growing on nitrogen and non-nitrogen fixing hosts and on CAM plants in the Namib Desert confirm partial heterotrophy. Oecologia 88: 457–462

Scott FM (1932) Some features of the anatomy of *Fouquieria splendens.* Am J Bot 19: 673–678

Scott FM (1935a) Contribution to the causal anatomy of the desert willow, *Chilopsis linearis.* Am J Bot 22: 332–343

Scott FM (1935b) The anatomy of *Cercidium torreyanum* and *Parkinsonia microphyllum.* Bull Torrey Bot Club 62: 33–41

Seddon G (1974) Xerophytes, xeromorphs and sclerophylls: the history of some concepts in ecology. Biol J Linn Soc 6: 65–87

Seely MK (1978) The Namib dune desert: an unusual ecosystem. J Arid Environ 1: 117–128

Sen DN (1968) Leafless *Euphorbia* in Rajasthan rocks, India. I. Ecological life history. Folia Geobot Phytotax 3: 1–15

Sen DN, Chawan DD (1972) Leafless *Euphorbia* in Rajasthan rocks, India. IV. Water relations of seedlings and adult plants. Vegetatio 24: 193–214

Sen DN, Rajpurohit KS (eds) (1982) Contributions to the biology of halophytes. Dr W Junk, The Hague

Sen DN, Sharma KD, Chawan DD (1971) Leafless *Euphorbia* in Rajasthan rocks. V. The organic acid metabolism of *E. caducifolia* Haines. New Phytol 70: 381–387

Sen DN, Chawan DD, Sharma KD (1972) Ecology of Indian Desert. V. On the water relations of *Salvadora* species. Flora 161: 463–471

Serrato Valenti G, Melne L, Modenesi P, Bozzini A (1989) Leaflet structure of *Prosopis tamarugo* Phil. (Leguminosae) and its water uptake. Phytomorphology 39: 181–188

Seybold A (1955) Beiträge zur Optik der Laubblätter. Beitr Biol Pflanz 31: 499–513

Shantz HL (1927) Drought resistance and soil moisture. Ecology 8: 145–157

Sharifi MR, Nilsen ET, Rundel PW (1982) Biomass and net primary production of *Prosopis glandulosa* (Fabaceae) in the Sonoran Desert of California. Am J Bot 69: 760–767

Sharifi MR, Nilsen ET, Virginia R, Rundel PW, Jarrell WM (1983) Phenological patterns of current season shoots of *Prosopis glandulosa* var. *torreyana* in the Sonoran Desert of southern California. Flora 173: 265–277

Sharma ML (1982) Aspects of salinity and water relations of Australian chenopods. In: Sen DN, Rajpurohit KS (eds) Contributions to the ecology of halophytes. Dr W Junk, The Hague, pp 155–172

Sheriff DW (1984) Epidermal transpiration and stomatal response to humidity: some hypotheses explored. Plant Cell Environ 7: 669–672

Shields LM (1950) Leaf xeromorphy as related to physiological and structural influences. Bot Rev 16: 399–447

Shields LM (1951a) Leaf xeromorphy in dicotyledons species from gypsum sand deposit. Am J Bot 38: 175–190

Shields LM (1951b) The involution mechanism in leaves of certain xeric grasses. Phytomorphology 1: 225–241

Shimony C, Fahn A (1968) Light- and electron-microscopical studies on the structure of salt glands of *Tamarix aphylla*. J Linn Soc Bot 60: 283–288

Shmida A (1985) Biogeography of the desert flora. In: Evenari M, Noy-Meir I, Goodall DW (eds) Hot deserts and arid shrublands. Ecosystems of the world, vol 12A. Elsevier, Amsterdam, pp 23–77

Shreve F (1924) Soil temperature as influenced by altitude and slope exposure. Ecology 5: 128–136

Shreve F (1951) Vegetation of the Sonoran Desert. Carnegie Institution Wash Pub 591, Washington, DC

Shreve F, Wiggins IL (1964) Vegetation and flora of the Sonoran Desert, 2 vols. Stanford University Press, Stanford

Shultz LM (1986) Comparative leaf anatomy of sagebrush (*Artemisia tridentata* complex): ecological considerations. In: McArthur ED, Welch BL (eds) Symposium on the biology of *Artemisia* and *Chrysothamnus*. Intermontane Research Station Pub, Ogden, pp 253–264

Smaoui MA (1971) Differentiation des trichomes chez *Atriplex halimus* L. Compt Rend Acad Sci Paris 273: 1268–1271

Smith BN, Turner BL (1975) Distribution of Kranz syndrome among Asteraceae. Am J Bot 62: 541–545

Smith GF, van Wyk AE (1992) Systematic leaf anatomy of selected genera of southern African Alooideae (Asphodelaceae). S Afr J Bot 58: 349–357

Smith JAC, Schulte PJ, Nobel PS (1987) Water flow and water storage in *Agave deserti*: osmotic implications of crassulacean acid metabolism. Plant Cell Environ 10: 639–648

Smith SD, Nobel PS (1986) Deserts. In: Baker NR, Long SP (eds) Photosynthesis in contrasting environments. Elsevier, Amsterdam, pp 13–62

Smith SD, Osmond CB (1987) Stem photosynthesis in a desert ephemeral, *Eriogonum inflatum*. Morphology, stomatal conductance and water-use efficiency in field populations. Oecologia 72: 533–541

Smith SD, Hartsock TL, Nobel PS (1983) Ecophysiology of *Yucca brevifolia*, an arborescent monocot of the Mojave Desert. Oecologia 60: 10–17

Smith SD, Didden-Zopfy B, Nobel PS (1984) High-temperature responses of North American cacti. Ecology 65: 643–651

Smith SD, Herr CA, Leary KL, Piorkowski JM (1995) Soil-plant water relations in a Mojave Desert mixed shrub community: a comparison of three geomorphic surfaces. J Arid Environ 29: 339–351

Smith SD, Monson RK, Anderson JE (1996) Physiological ecology of North American desert plants. Springer, Berlin Heidelberg New York

Smith WK (1978) Temperatures of desert plants: another perspective on the adaptability of leaf size. Science 201: 614–616

Smith WK, Nobel PS (1977a) Temperature and water relations for sun and shade leaves of a desert broadleaf, *Hyptis emoryi*. J Exp Bot 28: 169–183

Smith WK, Nobel PS (1977b) Influences of seasonal changes in leaf morphology on water-use efficiency for three desert broadleaf shrubs. Ecology 58: 1033–1043

Smith WK, Nobel PS (1978) Influence of irradiation, soil water potential, and leaf temperature on leaf morphology of a desert broadleaf, *Encelia farinosa* Gray (Compositae). Am J Bot 65: 429–432

Solbrig OT, Orians GH (1977) The adaptive characteristics of desert plants. Am Sci 65: 412–421

Spalding ES (1905) Mechanical adjustment of the sahuaro (*Cereus giganteus*) to varying quantities of stored water. Bull Torrey Bot Club 32: 57–68

Specht RL (1969) Comparison of the sclerophyllous vegetation characteristics of Mediterranean-type climates in France, California and southern Australia. I. Structure, morphology and succession. Aust J Bot 17: 277–292

Staples RC, Toenniessen GH (eds) (1984) Salinity tolerance in plants. Wiley-Interscience, New York

Steinbrinck C, Schinz H (1908) Über die anatomische Ursache der hydrochastischen Bewegungen der sog Jerichorosen und einiger anderen Wüstenpflanzen (*Anastatica, Odontospermum, Geigeria, Fagonia, Zygophyllum*). Flora 98: 471–502

Steudle E, Lüttge U, Zimmermann U (1975) Water relations of the epidermal bladder cells of the halophytic species *Mesembryanthemum crystallinum*: direct measurements of hydrostatic pressure and hydraulic conductivity. Planta 126: 229–246

Stewart GR, Press MC (1990) The physiology and biochemistry of parasitic angiosperms. Annu Rev Plant Physiol Mol Biol 41: 127–151

Stocker O (1970) Der Wasser- und Photosynthese-Haushalt von Wüstenpflanzen der mauretanischen Sahara. I. Regengrüne und immergrüne Bäume. Flora 159: 539–572

Stocker O (1971) Der Wasser- und Photosynthese-Haushalt von Wüstenpflanzen der mauretanischen Sahara. II. Wechselgrüne, Rutenzweig- und stammsukkulente Bäume. Flora 160: 446–494

Stocker O (1972) Der Wasser- und Photosynthese-Haushalt von Wüstenpflanzen der mauretanischen Sahara. III. Kleinsträucher, Stauden und Gräser. Flora 161: 46–110

Stocker O (1974a) Der Wasser- und Photosynthesehaushalt von Wüstenpflanzen der südalgerischen Sahara. I. Standorte und Versuchspflanzen. Flora 163: 46–88

Stocker O (1974b) Der Wasser- und Photosynthesehaushalt von Wüstenpflanzen der südalgerischen Sahara. II. Tagesserien. Flora 163: 89–142

Stone AM, Mason CT Jr (1979) A study of stem inflation in wild buckwheat, *Eriogonum inflatum*. Desert Plants 1: 77–81

Strain BR (1970) Field measurements of tissue water potential and carbon dioxide exchange in the desert shrubs *Prosopis julifera* and *Larrea divaricata*. Photosynthetica 4: 118–122

Strain BR, Chase VC (1966) Effect of past and prevailing temperatures on the carbon dioxide exchange capacities of some woody desert perennials. Ecology 47: 1043–1045

Strain BR, Johnson PJ (1963) Corticular photosynthesis and growth in *Populus tremuloides*. Ecology 44: 581–584

Sundberg MD (1985) Trends in distribution and size of stomata in desert plants. Desert Plants 7: 154–157

Sundberg MD (1986) A comparison of stomatal distribution and length in succulent and non-succulent desert plants. Phytomorphology 36: 53–66

Szarek SR, Ting IP (1975) Photosynthetic efficiency of CAM plants in relation to C_3 and C_4 plants. In: Marcelle R (ed) Environmental and biological control of photosynthesis. Dr W Junk, The Hague, pp 289–297

Szarek SR, Woodhouse RM (1976) Ecophysiological studies of Sonoran Desert plants. I. Diurnal photosynthesis patterns of *Ambrosia deltoidea* and *Olneya tesota*. Oecologia 26: 225–234

Szarek SR, Woodhouse RM (1977) Ecophysiological studies of Sonoran Desert plants. II. Seasonal photosynthesis patterns and primary production of *Ambrosia deltoidea* and *Olneya tesota*. Oecologia 28: 365–375

Szarek SR, Woodhouse RM (1978a) Ecophysiological studies of Sonoran Desert plants. III. The daily course of photosynthesis for *Acacia greggii* and *Cercidium microphyllum*. Oecologia 35: 285–294

Szarek SR, Woodhouse RM (1978b) Ecophysiological studies of Sonoran Desert Plants. IV. Seasonal photosynthetic capacities of *Acacia greggii* and *Cercidium microphyllum*. Oecologia 37: 221–229

Taiz L, Zeiger E (1991) Plant physiology. Benjamin/Cummings, Redwood City

Tanner V, Eller BM (1986) Epidermis structure and its significance for the optical properties of leaves of the Mesembryanthemaceae. J Plant Physiol 125: 285–294

Tevis L Jr (1958) A population of desert ephemerals germinated by less than one inch of rain. Ecology 38: 688–695

Thoday D (1933) The terminology of "xerophytism." J Ecol 21: 1–6

Thomas HH (1921) Some observations on plants in the Libyan Desert. J Ecol 9: 75–89

Thomson WW (1975) The structure and function of salt glands. In: Poljakoff-Mayber A, Gale J (eds) Plants in saline environments. Ecological studies, vol 15. Springer, Berlin Heidelberg New York, pp 118–146

Thomson WW, Berry WL, Liu IL (1969) Localization and secretion of salt by the salt glands of *Tamarix aphylla*. Proc Natl Acad Sci USA 63: 310–317

Thomson WW, Platt-Aloia K, Koller D (1979) Ultrastructure and development of the trichomes of *Larrea* (creosote bush). Bot Gaz 140: 249–260

Thorne RF (1982) The desert and other transmontane plant communities of southern California. Aliso 10: 219–257

Thorne RF (1992) An updated phylogenetic classification of the flowering plants. Aliso 13: 365–389

Thornthwaite CW (1948) An approach towards a rational classification of climate. Geogr Rev 21: 633–655

Ting IP, Gibbs M (eds) (1982) Crassulacean acid metabolism. American Society of Plant Physiologists, Rockville

Tissue DT, Nobel PS (1988) Parent-ramet connections in *Agave desert*: influence of carbohydrates on growth. Oecologia 75: 266–271

Tissue DT, Yakir D, Nobel PS (1991) Diel water movement between parenchyma and chlorenchyma of two desert CAM plants under dry and wet conditions. Plant Cell Environ 14: 407–413

Toft NL, Pearcy RW (1982) Gas exchange characteristics and temperature relations of two desert annuals: a comparison of a winter-active and a summer-active species. Oecologia 55: 170–177

Tomlinson PB (1961) Anatomy of the monocotyledons, vol 2. Palmae. Clarendon Press, Oxford

Trachtenberg S, Fahn A (1981) The mucilage cells of *Opuntia ficus-indica* (L.) Mill.—development, ultrastructure, and mucilage secretion. Bot Gaz 142: 206–213

Tschirch A (1882) Beiträge zu der Anatomie und dem Einrollungsmechanismus einiger Grasblätter. Pringsheim's Jahrb Wiss Bot 13(3): 1–26

Turner NC (1979) Drought resistance and adaptation to water deficits in crop plants. In: Mussell H, Staples RC (eds) Stress physiology of crop plants. Wiley-Interscience, New York, pp 343–372

Turner NC, Jones MM (1980) Turgor maintenance by osmotic adjustment: a review and evaluation. In: Turner NC, Kramer PJ (eds) Adaptation of plants to water and high temperature stress. Wiley-Interscience, New York, pp 87–103

Turner RM (1963) Growth in four species of Sonoran Desert trees. Ecology 44: 760–765

Turrell FM (1936) The area of the internal exposed surface of dicotyledon leaves. Am J Bot 23: 255–264

Ullmann I (1985) Tagesgänge von Transpiration und stomatärer Leitfähigkeit Sahelischer und Saharischer Akazien in der Trockenzeit. Flora 176: 383–409

Ullmann I, Lange OL, Ziegler H, Ehleringer JR, Schulze E-D, Cowan IR (1985) Diurnal courses of leaf conductance and transpiration of mistletoes and their hosts in central Australia. Oecologia 67: 577–587
Vankat JL (1989) Water stress in chaparral shrubs in summer-rain versus summer-drought climates—whither the mediterranean-type climate paradigm? In: Keeley JE (ed) The California chaparral: paradigms reexamined. Natural History Museum of Los Angeles Science Series 34, Los Angeles, pp 117–124
Vasek FC (1980) Creosote bush: long-lived clones in the Mojave Desert. Am J Bot 67: 246–255
Vischer W (1915) Experimentelle Beiträge zur Kenntnis der Jugend- und Folgeformen xerophiler Pflanzen. Flora 108: 1–72
Volkens G (1887) Die Flora der aegyptisch-arabischen Wüste auf Grundlage anatomisch-physiologischer Forschungen. Borntraeger, Berlin
von Willert DJ, Brinckmann E, Scheitler B, Schulze E-D, Thomas DA, Treichel S (1980) Ökophysiologische Untersuchungen an Pflanzen der Namib-Wüste. Naturwissenschaften 67: 21–28
von Willert DJ, Eller BM, Brinckmann E, Baasch R (1982) CO_2 gas exchange and transpiration of *Welwitschia mirabilis* Hook. fil. in the central Namib Desert. Oecologia 55: 21–29
von Willert DJ, Eller BM, Werger MJA, Brinckmann E, Ihlenfeldt H-D (1992) Life strategies of succulents in deserts: with special reference to the Namib Desert. Cambridge University Press, Cambridge
Wainwright CM (1977) Sun-tracking and related leaf movements in a desert lupine (*Lupinus arizonicus*). Am J Bot 64: 1032–1041
Waisel Y (1963) Ecotypic differentiation in the flora of Israel. III. Anatomical studies of some ecotype pairs. Bull Res Counc Isr Sec D11: 183–190
Waisel Y (1972) Biology of halophytes. Academic Press, New York
Wallace CS, Szarek SR (1981) Ecophysiological studies of Sonoran Desert plants. VII. Photosynthetic gas exchange of winter ephemerals from sun and shade environments. Oecologia 51: 321–333
Walter H (1986) The Namib Desert. In: Evenari M, Noy-Meir I, Goodall DW (eds) Hot deserts and arid shrublands. Ecosystems of the world, vol 12B. Elsevier, The Hague, pp 245–282
Walter H, Breckle S-W (1984) Ökologie der Erde, Bd 2. Spezielle Ökologie der tropischen und subtropischen Zonen. Gustav Fischer, Stuttgart
Warming E (1909) Oecology of plants: an introduction to the study of plant communities. Oxford University Press, Oxford
Weaver JE, Clements FE (1929) Plant ecology. McGraw-Hill, New York
Webber IE (1936) The woods of sclerophyllous and desert shrubs of California. Am J Bot 23: 181–188
Weber DJ, Rasmussen HP, Hess WM (1977) Electron microprobe analyses of salt distribution in the halophyte *Salicornia pacifica* var. *utahensis*. Can J Bot 55: 1516–1523
Weiglin C, Winter E (1991) Leaf structures of xerohalophytes from an East Jordanian salt pan. Flora 185: 405–426
Went FW (1948) Ecology of desert plants. I. Observations on germination in the Joshua Tree National Monument, California. Ecology 29: 242–253
Went FW (1949) Ecology of desert plants. II. The effect of rain and temperature on germination and growth. Ecology 30: 1–13
Went FW (1955) The ecology of desert plants. Sci Am 192(4): 68–75
Went FW (1971) Parallel evolution. Taxon 20: 197–226
Werk KS, Ehleringer J (1984) Non-random leaf orientation in *Lactuca serriola*. Plant Cell Environ 7: 81–87
Werk KS, Ehleringer J (1986) Field water relations of a compass plant, *Lactuca serriola* L. Plant Cell Environ 9: 681–683
Whittaker RH, Niering WA (1975) Vegetation of the Santa Catalina Mountains, Arizona. V. Biomass, production, and diversity along the elevation gradient. Ecology 56: 771–790
Wilkinson HP (1979) The plant surface (mainly leaf). In: Metcalfe CR, Chalk L (eds) Anatomy of the dicotyledons, vol 1, 2nd edn. Clarendon Press, Oxford, pp 97–165
Wilkinson RE (1966) Seasonal development of anatomical structures of saltcedar foliage. Bot Gaz 133: 73–77

Winkworth RE (1967) The composition of several arid spinifex grasslands of Central Australia in relation to rainfall, soil water relations, and nutrients. Aust J Bot 15: 107–130
Winter KC (1973) CO_2-Gaswechsel von an hohe Salinität adaptiertem *Mesembryanthemum crystallinum* bei Rückführung in glykisches Anzuchtmedium. Ber Dtsch Bot Ges 86: 467–476
Winter K (1981) C_4 plants of high biomass in arid regions of Asia—occurrence of C_4 photosynthesis in Chenopodiaceae and Polygonaceae from the Middle East and USSR. Oecologia 48: 100–106
Winter K, Schramm MJ (1986) Analysis of stomatal and nonstomatal components in the environmental control of CO_2 exchange in leaves of *Welwitschia mirabilis*. Plant Physiol 82: 173–178
Winter K, Troughton JH (1978) Photosynthetic pathways in plants of coastal and inland habitats of Israel and the Sinai. Flora 167: 1–34
Winter K, Kramer D, Troughton JH, Card KA, Pischer K (1977) C_4 pathway of photosynthesis in a member of the Polygonaceae: *Calligonum persicum* (Boiss. and Buhse) Boiss. Z Pflanzenphysiol 81: 341–346
Wood JG (1934) The physiology of xerophytism in Australian plants. The stomatal frequencies, transpiration and osmotic pressures of sclerophyll and tomentose-succulent leaved plants. J Ecol 22: 69–87
Woodhouse RM, Williams JG, Nobel PS (1983) Simulation of plant temperature and water loss by the desert succulent, *Agave deserti*. Oecologia 57: 291–297
Wooley JT (1964) Water relations of soybean leaf hairs. Agron J 56: 569–571
Wullstein LH, Pratt SA (1981) Scanning electron microscopy of rhizosheaths of *Oryzopsis hymenoides*. Am J Bot 68: 408–419
Zamski E, Ucko O, Koller D (1983) The mechanism of root contraction in *Gymnarrhena micranatha*, a desert plant. New Phytol 95: 29–35
Zeiger E (1983) The biology of stomatal guard cells. Annu Rev Plant Physiol 34: 441–475
Zemke E (1939) Anatomische Untersuchungen an Pflanzen der Namibwüste (Deutsch-Südwestafrika). Flora 33: 365–416
Ziegler H, Batanouny KH, Sankhla N, Vyas OP, Stichler W (1981) The photosynthetic pathway types of some desert plants from India, Saudi Arabia, Egypt, and Iraq. Oecologia 48: 93–99
Zimmermann MH (1983) Xylem structure and the ascent of sap. Springer, Berlin Heidelberg New York
Zohary M (1952) Ecological studies in the vegetation of the Near Eastern deserts. I. Environmental vegetation classes. Isr Exploration J Jerusalem 2: 210–215
Zohary M (1961) On hydro-ecological relations of the Near East desert vegetation. In: Plant-water relationships in arid and semi-arid conditions. Proc Madrid Symp, Arid Zone Res (UNESCO) 16: 199–212
Zohary M, Orshan G (1956) Ecological studies in the vegetation of the Near East deserts II. Wadi Araba. Vegetatio 7: 15–37

Taxonomic Index

Subject Index

MIX
Papier aus verantwortungsvollen Quellen
Paper from responsible sources
FSC® C105338

If you have any concerns about our products,
you can contact us on
ProductSafety@springernature.com

In case Publisher is established outside the EU,
the EU authorized representative is:
Springer Nature Customer Service Center GmbH
Europaplatz 3, 69115 Heidelberg, Germany

Printed by Libri Plureos GmbH
in Hamburg, Germany